The Black Death

Also by Philip Ziegler

THE
BLACK
DEATH

Philip Ziegler

HARPER**PERENNIAL** ● MODERN**CLASSICS**

NEW YORK ● LONDON ● TORONTO ● SYDNEY ● NEW DELHI ● AUCKLAND

HARPERPERENNIAL ◉ MODERNCLASSICS

First hardcover edition of this book was published in 1969 by the John Day Company, Inc.

P.S.™ is a trademark of HarperCollins Publishers.

HarperCollins books may be purchased for educational, business, or sales promotional use. For information please write: Special Markets Department, HarperCollins Publishers, 10 East 53rd Street, New York, NY 10022.

First Harper Torchbooks edition published 1971.

First Harper Perennial Modern Classics edition published 2009.

Library of Congress Cataloging-in-Publication Data is available upon request.

ISBN 978-0-06-171898-4

11 12 13   RRD   10 9 8 7 6 5 4 3 2

To Billy and Pierre

Contents

Preface

Though there may be controversy over its precise significance, no one would to-day deny that the Black Death was of the greatest economic and social importance as well as hideously dramatic in its progress. It is therefore surprising that, with the exception of Dr Coulton's whimsical monograph, no general study of the subject has appeared since Cardinal Gasquet wrote *The Great Pestilence* in 1893. In the meantime much new information has come to light and many of the dogmas accepted in the nineteenth century have been disproved or qualified.

I well understand the considerations which inhibit the academic historian from embarking on such a labour. Any general study must be either superficial or unwieldy: the first distasteful to the author, the second to the publisher and public. No one, too, can be expert in every aspect of this vast panorama and the spectacle of rival historians, each established in his fortress of specialised knowledge, waiting to destroy the unwary trespasser, is calculated to discourage even the most intrepid.

In Professor Elton's kindly if contemptuous phrase, I am an amateur and come to my task 'in a happy spirit of untrained enterprise.' I have read massively but lay no claim to the title of medievalist. This, I hope, may excuse me for rushing in where even an angel or a Professor of Medieval History would fear to tread. (To quote Professor Elton again—the angels may perhaps be forgiven if rather than tread themselves in those treacherous paths they prefer to bide their time and tread upon the fools instead.) This book contains virtually no original research. It is an attempt to synthesise in a single readable but reasonably comprehensive volume the records of the contemporary chroniclers and the work of later historians, in particular the great flood of PhD theses, each treating some tiny aspect of this enormous subject. And if this book should chance to provoke some academic historian—incensed by its inadequacy—into engaging in a major work of scholarship, then it will have served a useful purpose.

This therefore is not a book for the professional historian. I hope, however, that it may be of some use to him, if only as the solitary study which, in its bibliography and its references, has attempted to cover every serious contribution made by students of the Black Death. For this reason I have not dispensed with the paraphernalia of notes though I have exiled them to the end of the book. There is nothing in them to concern the general reader who does not wish to know the source of any particular piece of information.

Although I have dealt with the Black Death in England more thoroughly than elsewhere, I have tried also to give some indication of its origins and to sketch in the outline of its progress across Europe. The result is untidy but to confine myself to England or to the British Isles would have been to sacrifice all perspective in favour of a neat but narrow pattern.

A book of this nature poses many problems of construction. It would, in some ways, have been more logical to treat in isolation such subjects as the persecution of the Jews or the state of medical knowledge. On reflection however, and with some doubts, I have concluded that the book is little less lucid and considerably easier to read if such topics are dealt with as they seem naturally to arise in the course of the narrative.

Mr Richard Ollard, Mr Handasyde Buchanan and my brother, Mr Oliver Ziegler have read my manuscript with fortitude and made suggestions of great value. Dr Keele of the Wellcome Institute of Medical History was kind enough to read and criticise those sections relating to the nature and history of bubonic plague. Miss Barbara Dodwell, Reader in Medieval History at the University of Reading, has corrected many of my blunders and pointed out a variety of ways by which the book might be improved.

Any writer who leans as heavily as I on the work of other people must stagger under an almost intolerable burden of gratitude. I acknowledge with pleasure and admiration my debt to all those whose publications I have made use of in this book. Two names, above all, I would like to single out: the late Professor Hamilton Thompson, whose pioneer work in this field has made so vast a contribution to the labour of all subsequent historians and Dr Elizabeth Carpentier, of the École des Hautes Études in Paris, whose wit and scholarship to-day illumine the same subject.

ONE

Origins and Nature

It must have been at some time during 1346 that word first reached Europe of strange and tragic happenings far away in the East. Even in this age of easy travel and rapid spread of news, calamities in China tend to be accepted in the Occident with the polite but detached regret reserved for something infinitely remote. In the fourteenth century, Cathay was a never-never-land; unheard of except by the more sophisticated and, even to them, a place of mystery which only a few merchants had visited and about which little was known. No story, however horrific, would seem altogether implausible if it came from such a source; but equally no medieval savant or merchant would have conceived that what happened so far away could have any possible relevance to his own existence. The travellers' tales were received with awed credulity but gave rise to no alarm.

Certainly things seemed to have gone badly wrong. An imposing series of disasters studded the history of the previous years.[1] In 1333 parching drought with consequential famine had ravaged the plains watered by the rivers Kiang and Hoai. Then had come floods in which four hundred thousand were said to have died, as a result of which, presumably, the mountain Tsincheou 'fell in,' causing great chasms in the earth. In 1334 there was drought in Houkouang and Honan followed by swarms of locusts, famine and pestilence. An earthquake in the mountains of Ki-Ming-Chan formed a lake more than a hundred leagues in circumference. In Tche the dead were believed to number more than five million. Earthquakes and floods continued from 1337 to 1345; locusts had never been so destructive; there was 'subterraneous thunder' in Canton.

But these were mere curtain-raisers for the real calamity. Several contemporary accounts exist of the earliest days of the

Black Death, so similar in detail as to suggest that they may well
have come from the same source. Almost the only man known to
have been at or near the spot, Ibn-Bātuta, 'The Traveller,' is dis-
appointingly reticent.[2] An anonymous Flemish cleric, on the other
hand, was fortunately unfettered by the restrictions imposed on
those who have actually seen what they describe. Basing himself
on a letter from a friend in the papal curia at Avignon he re-
counted how: 'in the East, hard by Greater India, in a certain
province, horrors and unheard of tempests overwhelmed the
whole province for the space of three days. On the first day there
was a rain of frogs, serpents, lizards, scorpions, and many veno-
mous beasts of that sort. On the second, thunder was heard, and
lightning and sheets of fire fell upon the earth, mingled with hail
stones of marvellous size; which slew almost all, from the greatest
even to the least. On the third day there fell fire from heaven and
stinking smoke, which slew all that were left of men and beasts,
and burned up all the cities and towns in those parts. By these
tempests the whole province was infected; and it is conjectured
that, through the foul blast of wind that came from the South, the
whole seashore and surrounding lands were infected, and are
waxing more and more poisonous from day to day . . .'[3]

This concept of a corrupted atmosphere, visible in the form of
mist or smoke, drifting across the world and overwhelming all
whom it encountered, was one of the main assumptions on which
the physicians of the Middle Ages based their efforts to check the
plague. For one chronicler the substance of the cloud was more
steam than smoke.[4] Its origin was to be found in a war which had
taken place between the sea and the sun in the Indian Ocean. The
waters of the ocean were drawn up as a vapour so corrupted by
the multitude of dead and rotting fish that the sun was quite
unable to consume it nor could it fall again as healthy rain. So it
drifted away, an evil, noxious mist, contaminating all it touched.
For the Chronicler of Este, however, this cloud of death owed
nothing to the sea:

'Between Cathay and Persia there rained a vast rain of fire;
falling in flakes like snow and burning up mountains and plains
and other lands, with men and women; and then arose vast
masses of smoke; and whosoever beheld this died within the
space of half a day; and likewise any man or woman who looked
upon those who had seen this . . .'[5]

By the end of 1346, therefore, it was widely known, at least in the major European seaports, that a plague of unparalleled fury was raging in the East. Fearful rumours were heard of the disease's progress: 'India was depopulated, Tartary, Mesopotamia, Syria, Armenia were covered with dead bodies; the Kurds fled in vain to the mountains. In Caramania and Caesarea none were left alive . . .'[6] But still it does not seem to have occurred to anyone that the plague might one day strike at Europe.

The most circumstantial account of how the disease made this fatal leap comes from Gabriel de Mussis. At one time, indeed, it was thought that the writer had himself visited Asia Minor and had been a passenger on the ship which carried the plague to Europe.[7] A subsequent editor, however, has reluctantly but decisively established that de Mussis, during the critical period, never stirred from his native town of Piacenza.[8]

De Mussis stated that the plague settled in the Tartar lands of Asia Minor in 1346. According to Vernadsky it left eighty-five thousand dead in the Crimea alone.[9] Whether coincidentally or because they made the conventional medieval assumption that some human agency, preferably in the form of an already unpopular minority group, must be responsible for their sufferings, the Tartars decided to attack the Christian merchants in the vicinity. A street brawl, in which one of the locals was killed, seems to have provided the excuse for what was probably a premeditated campaign. The Tartars set on a Genoese trading station in the city of Tana and chased the merchants to their redoubt at Caffa, now Feodosia, a town on the Crimean coast which the Genoese had built and fortified as a base from which to trade with the Eastern hinterland. The Tartar army settled down outside the walls and prepared to bombard the city into submission.[10]

Their plans were disastrously disturbed by the plague which was soon taking heavy toll of the besiegers. 'Fatigued, stupefied and amazed,' they decided to call off the operation. First, however, they felt it was only fair that the Christians should be given a taste of the agony which the investing force had been suffering. They used their giant catapults to lob over the walls the corpses of the victims in the hope that this would spread the disease within the city. As fast as the rotting bodies arrived in their midst the Genoese carried them through the town and dropped them in the sea. But few places are so vulnerable to disease as a besieged city and it was not long before the plague was as active within the city as without. Such inhabitants as did not rapidly succumb realised that, even if they survived the plague, they would be far too few to resist a fresh Tartar onslaught. They took to their galleys and fled from the Black Sea towards the Mediterranean. With them travelled the plague.

Though it is certain that this can not have been the only, and probable that it was not even the earliest route by which the plague arrived in Europe there is no reason to doubt that de Mussis's story is true in essentials. One of the main trade routes by which the spices and silks from the East reached the European market was by way of Baghdad and thence along the Tigris and through Armenia to the entrepot stations of the Italian merchants in the Crimea. Nothing is more likely than that the plague should travel with the great caravans and spread itself among the Tartars of the Crimea, the 'hyperborean Scythians' who, in the opinion of the Byzantine Emperor, John Cantacuzenos, were the first victims of the epidemic.[11]

According to one chronicler: 'This plague on these accursed galleys was a punishment from God, since these same galleys had helped the Turks and Saracens to take the city of Romanais which belonged to the Christians and broke down the walls and slew the Christians as though they were cattle or worse; and the Genoese wrought far more slaughter and cruelty on the Christians than even the Saracens had done.' As with other instances of Divine retribution, the punishment seems to have been strikingly promiscuous. For even though it is not necessary to believe, with the Chronicler of Este, that these ill-fated galleys, with the crews dying at their oars, somehow contrived to spread the plague to 'Constantinople, Messina, Sicily, Sardinia, Genoa, Marseilles and many other places,' it is likely that infection was carried at least to Genoa, Venice and Messina by galleys from Eastern ports.[12]

The inhabitants reacted violently when they realised the cargo that their visitors were bringing with them. They sought to drive the danger away and, in so doing, ensured that it spread more rapidly. 'In January of the year 1348,' wrote a Flemish chronicler, 'three galleys put in at Genoa, driven by a fierce wind from the East, horribly infected and laden with a variety of spices and other valuable goods. When the inhabitants of Genoa learnt this, and saw how suddenly and irremediably they infected other people, they were driven forth from that port by burning arrows and divers engines of war; for no man dared touch them; nor was any man able to trade with them, for if he did he would be sure to die forthwith. Thus, they were scattered from port to port . . .'[13]

But by the time that the Genoese authorities reacted, it was too late. The infection was ashore and nothing was to stop it. By the spring of 1348 the Black Death had taken a firm grasp in Sicily and on the mainland.

At this point, with the plague poised to strike into the heart of Europe, it seems appropriate to pause and consider what the epidemic really was and how far it was something entirely new. To-day there is little mystery left about the origins and nature of the Black Death; a few points remain to be clarified but all the essential facts are known. But in the Middle Ages the plague was not only all-destroying, it was totally incomprehensible. Medieval man was equipped with no form of defence—social, medical or psychological—against a violent epidemic of this magnitude. His baffled and terrified helplessness in the face of disaster will be above all the theme of this book.

One of the minor mysteries which does still persist over the Black Death is the genesis of its name. The traditional belief is that it was so called because the putrefying flesh of the victims blackened in the final hours before death supervened. The trouble about this otherwise plausible theory is that no such phenomenon occurred. It is true that, in cases of septicaemic plague, small black or purple blotches formed on the bodies of the sick and this symptom must have made a vivid impression on beholders. But if the name of the epidemic had been derived primarily from the appearance of its victims, one would have expected it to have been used at the time. Of this there is no evidence. Indeed, it seems that such a title was not generally heard until the eighteenth century, though similar expressions had often been applied to other epidemics in the past.[14] The first recorded use of the term for the epidemic of 1348 is in a reference to the *swarta döden* in Sweden in 1555. About fifty years later it emerged in Denmark as the *sorte død*.[15] Cardinal

Gasquet believed that, in England at least, the name began to
be used sometime after 1665 to distinguish the fourteenth-
century epidemic from the 'Great Plague' which ravaged Carolean
London.[16]

The fact that the title 'Black Death' was not used by contemporaries similarly makes it hard to credit those other explana-
tions which attributed the name to a black comet seen before the
arrival of the epidemic, to the number of people who were
thrown into mourning as a result of the high mortality[17] or to the
popular images of the plague as a man on a black horse or as a
black giant striding across the countryside.[18]

The most likely explanation seems to be that it originally
stemmed from an over-literal translation into the Scandinavian
or the English of the Latin *pestis atra* or *atra mors*. Even in the
fourteenth century the word 'atra' could connote 'dreadful' or
'terrible' as well as 'black.' But once the mistranslation had been
established then all the other reasons for associating 'Black' with
'Death' must have contributed to give it general currency. In
France it was once called the *morte bleue*. The superior dreadfulness of the accepted phrase is obvious and to-day no other
style would be acceptable.

Contemporary records are remarkably consistent in their descriptions of the physical appearance of the disease. The most
commonly noted symptom is, of course, also the most dramatic;
the buboes or boils, sometimes also described as knobs, kernels,
biles, blaines, blisters, pimples or wheals, which are the invariable
concomitants of bubonic plague. Boccaccio's description will
do for all the rest:

'. . . in men and women alike it first betrayed itself by the
emergence of certain tumours in the groin or the armpits, some
of which grew as large as a common apple, others as an egg, some
more, some less, which the common folk called gavocciolo. From
the two said parts of the body this deadly gavocciolo soon began
to propagate and spread itself in all directions indifferently; after
which the form of the malady began to change, black spots or
livid making their appearance in many cases on the arm or the
thigh or elsewhere, now few and large, now minute and numerous. And as the gavocciolo had been and still was an infallible
token of approaching death, such also were these spots on whom-
soever they shewed themselves . . .'[19]

Medically the only questionable detail in this account is the reference to the bubo as an 'infallible token of approaching death.' Other contemporary records[20] as well as observation of subsequent epidemics show that it was by no means unheard of for the buboes to discharge and the patient recover. But certainly this happened in a very small minority of cases. To most of its victims the bubo meant inevitable death and it would not be surprising if Boccaccio had never heard of an instance to the contrary.

It was Gui de Chauliac, physician to the Papal Court at Avignon, who saw most clearly that these buboes were by no means an invariable symptom and that a distinct, still more violent variant of the plague existed.[21] 'The mortality . . . lasted seven months,' he wrote. 'It was of two types. The first lasted two months, with continuous fever and spitting of blood, and from this one died in three days. The second lasted for the rest of the period, also with continuous fever but with apostumes and carbuncles on the external parts, principally on the armpits and groin. From this one died in five days.'

The first form, de Chauliac had no doubt, was the more deadly. Even those doctors who failed to perceive the significance of the different symptoms, associated the coughing of blood with certain death: '. . . men suffer in their lungs and breathing and whoever have these corrupted, or even slightly attacked, cannot by any means escape nor live beyond two days.'

The question of how long the sick could be expected to survive caused much confusion to the contemporary chroniclers; confusion that could never be cleared up because of their failure to identify the second and, as we now know, the third distinct form of the plague. Most reports agreed with Boccaccio that, in those cases where there were only buboes, death was likely to come in five or six days but that, when there was coughing of blood, either by itself or as an additional symptom, the course of the disease was more rapid and the patient died within two or three days. But there were other, by no means infrequent references to the disease killing almost instantaneously or within a few hours. Geoffrey the Baker wrote[22] of people who went peacefully to bed and were dead the next morning, while Simon of Covino described priests or doctors who 'were seized by the plague whilst administering spiritual aid; and, often by a single touch, or a single breath of the plague-stricken, perished even before the sick person they had come to assist.'[23]

Through almost every account breathes the revulsion as well as the fear which the plague inspired in all who encountered it. Disease rarely respects human dignity and beauty but the Black Death seemed peculiarly well equipped to degrade and humiliate its victims. Everything about it was disgusting, so that the sick became objects more of detestation than of pity: '... all the matter which exuded from their bodies let off an unbearable stench; sweat, excrement, spittle, breath, so foetid as to be overpowering; urine turbid, thick, black or red . . .'[24]

All these phenomena were observed with horrified accuracy by contemporary writers and reported with care and objectivity. Little or no effort was made, however, to explain them logically or to work them into a coherent pattern; the background of knowledge against which such an attempt could have been made was woefully inadequate and the will to try rarely present. One subject which proved something of an exception was the problem of how the disease passed from man to man and country to country. To this much thought was given and many esoteric theories were put forward. Fundamentally there were two, by no means mutually exclusive schools of thought: those who believed in person-to-person infection and those who pinned their faith in the existence of a 'miasma' or poison cloud.

The medieval doctor was confronted with a situation where a large number of people died suddenly and inexplicably in a given area. This zone of mortality shifted constantly but gradually, conquering new territory and abandoning old. Any rational being, faced with such a phenomenon but totally unversed in medical lore, would be likely to arrive at the same explanation. There must be some vicious property in the air itself which travelled slowly from place to place, borne by the wind or impelled by its own mysterious volition. There were many different points of view as to the nature of this airborne menace, its origin, its physical appearance. But almost every fourteenth-century savant or doctor took it for granted that the corruption of the atmosphere was a prime cause of the Black Death.

Debate generally centred on the degree of harm which this corruption might do to the atmosphere which it affected. Ibn Khātimah, the Arab philosopher and physician from Granada, took an extreme line when he argued that corruption might in certain cases be absolute; that is to say that the very nature of the air might be permanently changed by putrefaction.[25] In such air no light would burn, still less could a man hope to live. Such conditions existed only in the very heart of a plague-ridden area, all around was to be found a zone of partial corruption where the danger of death, though still great, was no longer inescapable. A change in the composition of the air might be caused by the movement of the stars or the putrid fumes of decaying matter; in the case of the Black Death, however, Ibn Khātimah believed that the cause was to be found in the vagaries of the weather over the previous few years.

Not many people agreed with the theory that the air could be altered in its very composition. Even Ibn Khātimah's colleague and friend, Ibn al Khatīb, could not accept that there was more than a temporary poisoning caused by the addition of something noxious to the atmosphere.[26] What that something was and where it came from was similarly a topic for hot debate. Alfonso of Cordova, like most men of learning at the time, considered that movements of the planets probably started the mischief but held that, if the poisoning of the atmosphere went on for any length of time, then some human agency must be behind it:

'since air can also be infected artificially, as when a certain confection is prepared in a glass flask, and when it is well fermented, the person who wishes to do that evil waits till there is a strong, slow wind from some region of the world, then goes into the wind and rests his flask against some rocks opposite the city or town which he wishes to infect and, making a wide detour and going further into the wind lest the vapours infect him, pulls his flask violently against the rocks. When it breaks the vapour pours out and is dispersed in the air, and whoever it touches will die . . .'[27]

The anonymous author of another plague tractate rejected all such fantasies of medieval gas warfare and made the great earthquakes of 1347 the villains of the piece.[28] Even before this date, pressure had been building up underground and a few noxious currents of air which had escaped had started minor epidemics. But now poisonous fumes poured forth through cracks in the earth, escaped into the atmosphere and drifted around the face of Europe killing all who encountered them.

It is curious that, though every doctor paid lip-service to the teachings of Galen, the relatively prosaic explanation of the corruption of the atmosphere which he had advanced several hundred years before was almost ignored by commentators of the Black Death. Infection, he propounded, arose mainly from 'Inspiration of air infected with a putrid exhalation. The beginning of the putrescence may be a multitude of unburned corpses, as may happen in war; or the exhalations of marshes and ponds in the summer . . .' Perhaps the monstrous dimensions of the disaster which overtook Europe in the fourteenth century forced its victims to seek some proportionately monstrous explanation.

But the idea that the disease might be passed directly from man to man was not ruled out by belief in a corrupted atmosphere. A few, mainly among the Arabs, rejected the possibility of infection on religious grounds but for most people the evidence of their own eyes was too strong. Some effort was made to establish a link between the two theories, as by those who argued that a victim of the plague might radiate infection in his immediate vicinity by generating a form of personal and highly localised miasma which he carried, like a halo, around his head.[29] But such refinements of logic were not much considered and, in general, people were content to note that the disease could pass from victim to victim with terrifying speed and did not worry too much about the philosophical or scientific basis for such a phenomenon. The evidence was overwhelming. It is noteworthy that it was an Arab, Ibn al Khatīb, who defied his religion's teaching and stated flatly: 'The existence of infection is firmly established by experience, research, mental perception, autopsy and authentic knowledge of fact . . .'

It does indeed appear that, to the medieval mind, the speed with which the Black Death passed from man to man was its most alarming feature. 'The contagious nature of the disease,' wrote one chronicler,[30] 'is indeed the most terrible of all the terrors, for when anyone who is infected by it dies, all who see him in his sickness, or visit him, or do any business with him, or even carry him to the grave, quickly follow him thither, and there is no known means of protection.'

Boccaccio was particularly struck by the perils of infection: 'Moreover, the virulence of the pest was the greater by reason that intercourse was apt to convey it from the sick to the whole, just as fire devours things dry or greasy when they are brought close to it. Nay, the evil went yet further, for not merely by speech or association with the sick was the malady communicated to the healthy with consequent peril of common death, but any that touched the clothes of the sick or aught else that had been touched or used by them, seemed thereby to contract the disease.'

Boccaccio himself claims to have seen a couple of pigs in the street digging their snouts into the rags of a poor man who had just died. They smelt them, tossed them to and fro between their teeth, then almost immediately ran round and round, and, without more ado, tumbled dead to the ground. The inevitable result of this well-founded if somewhat exaggerated terror of infection was that the victims of the plague more often than not found themselves abandoned to their fate and even those who had endured some slight brush with the disease were likely to be shunned by their fellow-men. Security lay only in total isolation. If this were impracticable then at least contact with the plague-struck should be avoided.

It was quickly realised that there was no need to touch a sick man to be infected. Most people believed that the disease passed by breath but other theories existed. Looks, according to a physician from Montpellier, could kill.[31] 'Instantaneous death occurs when the aerial spirit escaping from the eyes of the sick man strikes the eyes of a healthy person standing near and looking at the sick, especially when the latter are in agony; for then the poisonous nature of that member passes from one to the other, killing the other.' But swift and terrible though the infection might be, it was also evident that it varied in its dreadfulness from place to place and time to time. On one occasion a whole community would be obliterated, on another there would only be one or two victims and the rest would survive unscathed; here a family would die within twenty-four hours, there the father would die one day, a child three weeks later, another child after a month and then there would be no further victim. In general this was accepted apathetically as yet another of those inexplicable phenomena of which the Black Death was composed. A few doctors noted that the infection seemed more virulent where there was spitting of blood. Only Gui de Chauliac went on to deduce that, of the two forms of the disease which were apparent, one was notably more infectious than the other.

Enjoying as we do the immense superiority of a generation which has devised means of mass destruction more effective even than those inflicted by nature on our ancestors, it is easy and tempting to deride their inability to understand the calamity which had overtaken them. It would, perhaps, be more fitting to wonder at the courage and wisdom of men like Gui de Chauliac who saw their civilisation apparently doomed by a hideous and inexplicable calamity and could still observe its development with scientific objectivity, draw reasonable deductions about its habits and likely course and do their best to curb its ravages. It is also sobering to reflect that only within the last century have we learned enough to detect the origins and plot the course of the epidemics and that, even to-day, quick and expensive action is necessary if they are to be checked before they do great damage. It is much less than a hundred years since the sophisticated and immensely learned Dr Creighton concluded positively that the source of the Black Death lay in the mounds of dead left unburied by the successive disasters which had overtaken China. He invoked cadaveric poisoning as the reason for the high death rate among priests and monks: priests tended to live near the village churchyard while: 'Within the monastery walls were buried not only generations of monks, but often the bodies of princes, of notables of the surrounding country, and of great ecclesiastics.'[32]

To-day we smile politely at Dr Creighton's blunders; it is reasonable to wonder whether a hundred years from now the theories of to-day may not seem equally ridiculous. On the whole it is unlikely that they will. Undoubtedly further discoveries will be made, dark corners illuminated, concepts amended or refined. But the techniques of scientific investigation are now sufficiently evolved to have established as a fact the main elements of the Black Death and to explain authoritatively the cycle of its activity.[33]

That the Black Death, in its original form, was bubonic plague has been commonly accepted for many years. Bubonic plague is endemic to certain remote areas of the world; those which have been identified with reasonable certainty are Uganda,

Western Arabia, Kurdistan, Northern India and the Gobi Desert. From time to time it erupts there in the form of minor, localised epidemics. Far more rarely it breaks its bounds and surges forth as one of the great pandemics. Unlike influenza, bubonic plague in such a mood moves slowly, taking ten years or more to run its course across the world. When it comes, it comes to stay. The high mortality of its initial impact is followed by a long period in which it lies endemic, a period interspersed with occasional epidemics which gradually die away in frequency and violence. Finally, perhaps several hundred years after the original outbreak, the plague vanishes.

Three such pandemics have been recorded. The first, beginning in Arabia, reached Egypt in the year 542. It ravaged and perhaps even fatally weakened the Roman Empire of Justinian and moved on across Europe to England, where it was known as the Plague of Cadwalader's Time, and Ireland, which it laid waste in 664. The second pandemic was that of the Black Death. One of its parting flourishes was the Great Plague of London in 1665; it seems to have died out in the seventeenth century. Finally came the pandemic which started in 1892 in Yunnan and reached Bombay in 1896. In India alone it is believed to have killed some six million people. It made a brief and mercifully unsuccessful foray into Suffolk in 1910, finding only a handful of victims. Quite recently it has made itself felt in the Azores and parts of South America. In many parts of the world it has still to run its course.

Though on present evidence it is impossible to be categoric about the origins of the medieval pandemic, recent investigations by the Russian archaeologist Chwolson near Lake Issyk-Koul in the district of Semiriechinsk in Central Asia show that there was an abnormally high death rate in 1338 and 1339. Nestorian memorial stones attribute the deaths to plague.[31] Given the later course of the disease and the fact that this area is in the heart of one of the zones in which bubonic plague lies endemic, Dr Pollitzer, probably the leading authority on the subject, has concluded that this was almost certainly the cradle of the Black Death.[35] From thence it spread out, eastwards into China, south to India and west to reach the Crimea some eight years later.

In this remote fastness, since recorded history, the bacillus *Pasteurella Pestis* has lingered on, finding its home either in the bloodstream of an animal or the stomach of a flea. The flea normally favoured is *Xenopsylla Cheopsis*, familiarly *X. Cheopsis*, an insect which, in its turn, chooses ideally to reside in the hair of some rodent. One can only guess which rodent was most readily to be found near Lake Issyk-Koul in 1338 but the experience of later epidemics points to the tarbagan or Manchurian marmot, a beguiling squirrel-like creature much hunted for its skin. The jerboa and the suslik probably also played their part and, of course, the rat too, though the latter's main role was not to come till the disease was on the move.

To disturb the tranquil and largely harmless existence of *Pasteurella Pestis* something had to happen to make the rodents leave their homes. With them, inevitably, would travel their attendant fleas and, within the fleas, a cargo of deadly parasites. We are unlikely ever to know exactly what it was which caused this particular rodent migration. Such evidence as survives suggests that they were driven away by floods but, on other occasions, prolonged droughts have provided the necessary incentive or it could simply have been that an increase in the rodent population put too great a strain on the available supplies of food. At all events a massive exodus took place and it was above all *rattus rattus*, the tough, nimble, by nature vagabond, black rat which made the move.

Without disputing the importance of the rat as a carrier of plague, Professor Jorge has suggested that its role, except in the earliest stages of an epidemic, is inessential, and that the lack of references to it in contemporary accounts of the Black Death indicates that the infection was mainly dependent on other means of transport.[36] He believed that *pulex irritans*, the flea which preys above all on human beings, was perfectly capable of carrying the plague direct from man to man without the intervention of an infected rat. Medically this is doubtful. There is no need to eliminate *pulex irritans* altogether as an extra factor but its capacity to drink in sufficient plague bacilli from one person so as to be able to implant a fatal dose in the next has been much questioned. Colonel MacArthur has recorded that, in blood cultures made from fatal cases of bubonic plague, he found 'bacilli so sparse that theoretically one could have fed twenty thousand fleas on such a case and yet have infected none.'[37]

There is certainly no doubt that the rapid spread of bubonic plague was greatly helped by the presence of infected rats. Nor was there any shortage of rats. By the middle of the fourteenth century they abounded in Europe, probably having been imported originally in the boats of the returning Crusaders. Their role was unobtrusive and, since there is no particular reason why contemporaries should have commented on their activities, their absence from the chronicles casts no doubt on their existence. Dead rats no doubt littered the streets and houses but this would hardly have seemed worthy of attention at a time when dead human beings were so much more conspicuous.

But though the rat helps greatly in the spread of bubonic plague, Professor Jorge is right in his contention that it is not essential. The Plague Research Commission of 1910 commented '. . . the transference of infected rats and fleas in merchandise or, in the case of fleas, on the body of a human being, must be considered.'[38] It has been, and Dr Hirst has shown that the adage 'No ship rats, no plague' is palpably untrue.[39] X. Cheopsis, in ideal conditions, can live for a month away from its host. Travelling with a cargo of grain or in a bale of cloth it could easily journey hundreds of miles without a rat. There is one substantiated case of a flea surviving unfed for six months in a rat burrow. The absence of rats, therefore, was far from a guarantee that bubonic plague could never strike.

The symptoms of bubonic plague as known to-day coincide precisely with those described by the medieval chroniclers. The 'swollen and dropsical mass of inflamed lymphatic glands' known as the bubo is the classic sign. Sometimes this is the size of an almond, sometimes of an orange; usually it is found in the groin but it may also grow in the armpit or, occasionally, on the neck. Equally familiar are the dusky stains or blotches caused by subcutaneous haemorrhages and the intoxication of the nervous system: 'In Provence a man climbed on to the roof of his house and threw down the tiles into the street. Another executed a mad, grotesque dance on the roof . . .'[40] Modern medical experience suggests that, if the bubo breaks down and suppurates within a week, the victim will probably survive; few medieval doctors would have expected their patient to endure more than four or five days of the agonising pain which accompanies the boil. But otherwise the cases observed by Boccaccio or Simon of Covino could be found in half a dozen plague centres to-day.

But though bubonic plague was the first and most conspicuous

form taken by the Black Death, a variant known as primary
pneumonic or pulmonary plague was more lethal. In the epidemics of the late nineteenth century, when methods of treatment
were remarkably little more sophisticated than in the Middle
Ages, between sixty and ninety percent of those who caught
bubonic plague could expect to die. In the case of pneumonic
plague recovery was virtually unknown. Bubonic plague would
generally take between four days and a week to kill; in the Manchurian epidemic of 1921 the expectation of life of the victims of
pneumonic plague was a mere 1.8 days. Finally, bubonic plague
is one of the less infectious epidemic diseases; the breath is not
affected and the patient usually dead or recovered before enough
bacilli have accumulated in the blood to make it a source of infective material for the flea.[41] Pneumonic plague is perhaps the
most infectious; it attacks the lungs so that there is coughing of
blood and the plague bacilli are sprayed out into the air every
time that the patient exhales.

Hirst has remarked[42] that, if it were not known that they had a
common origin and were linked by intermediate types, true
pneumonic and uncomplicated bubonic plague would seem to be
different diseases. The link between the two is to be found in an
attack of bubonic plague during which the victim also develops
pneumonia. This compound, though extremely dangerous to the
victim, is not usually infectious. Yet, in certain cases, it may become so. The main outstanding problem of the Black Death, or
indeed of the plague in any era, is what the factors are which
make this happen, what it is which provokes an epidemic of the
air-borne pneumonic variant of the disease.

'Where the fourteenth-century plague is said to differ from later
experience is that in its quite slow extension across Europe it
seemed to change as the season of the year changed from pneumonic to bubonic, and then from bubonic to pneumonic without
discontinuity.'[43] The medieval doctor can hardly be blamed for
finding the process incomprehensible. But if he had understood it
he would even then not have mastered the full story. For there
would still have remained unexplained those cases, already mentioned, in which a man would die within a few hours or go to
bed in the best of health and never wake in the morning.

There seems no doubt that a third element in the Black Death,
septicaemic plague, was here at work. This, like bubonic plague,

is insect borne. The distinction is that the brunt of the infection falls on the bloodstream which, within an hour or two, is swarming with plague bacilli. The victim is dead long before buboes have had time to form. It is in this form of plague that *pulex irritans*, the man-borne flea, has a chance to operate. So rich in bacilli is the blood of a sick man that the flea can easily infect itself and carry on the disease to a new prey without the need of a rat to provide fresh sources of infection. Septicaemic plague must have been the rarest of the three interwoven diseases which composed the Black Death but it was certainly as lethal as its pneumonic cousin and it introduced yet another means by which the plague could settle itself in a new area and spread hungrily among the inhabitants.

TWO

The State of Europe

In a book of this scope it would be over-ambitious to attempt any serious analysis of the economic and social state of Europe in the middle of the fourteenth century. Something however must be said; for the circumstances of the continent and the physical and mental condition of its inhabitants, are factors of the utmost importance when considering the impact of the Black Death. 'The plague of the fourteenth century,' wrote Michon,[1] 'was no different to those which preceded or which followed it. It killed more people, not because of its nature, but because of the conditions of suffering and servitude in which it surprised its victims.' No one who has studied the devastating blows which the Black Death struck against rich and poor, young and old, strong and weak, can accept that this was just another epidemic like any other. But Michon's assertion is not, for this reason, to be dismissed as idle rhetoric.

During the eleventh, and even more the twelfth and first half of the thirteenth centuries Europe enjoyed a period of massive and almost unbroken economic growth. Some historians have recently questioned whether, in England at least, the Golden Age of the 'high' Middle Ages was in fact so spectacularly prosperous as has been generally believed.[2] Of course, sectors of the economy can be identified which lagged behind the rest and certain areas fared less well than others. But on the whole what Professor Nabholz described as 'the astonishing uniformity of medieval conditions throughout the whole region'[3] ensured that the boom was general and that no part of Europe was left out altogether.

In the two centuries preceding the middle of the thirteenth century the face of Europe was changed, and changed vastly for the better. The Crusades syphoned off much of the belligerent tendencies of the inhabitants and the period was one of comparative calm. The peasantry throve in unaccustomed security; Froissart wrote appreciatively of '*le pays gras et plentureux de toutes choses . . . les maisons pleines de toutes richesses . . .*' Land in the valleys of the Rhine and the Moselle was worth seventeen times as much at the end of the thirteenth century as it had been at the start of the tenth, yet the old customary rents remained substantially unchanged.[4] Colonisation, that is to say the capture of virgin lands from hills, fens and forests, went on apace. By 1300, in Central and Western Europe, the amount of land under cultivation had reached a point not to be matched for another five hundred years.

The primary driving force behind the new colonisation was, of course, the pressure of population on existing resources. By the middle of the thirteenth century Europe was becoming uncomfortably over-crowded. The density of population around Pistoia was thirty-eight per square kilometre—crowded by the standards of any rural area though by no means unusual in medieval Tuscany. The province probably had a population of some 1.18 million, a total which was not to be reached again until well into the nineteenth century. The population had grown rapidly since the middle of the eleventh century; production of food had grown too but at nothing approaching the same rate. Nor did it seem that medieval techniques of agriculture were far enough advanced for the gap between demand and supply to do anything but widen. The Tuscan peasant, who had never lived far above the subsistence level, now found that he was near to falling below it.

Tuscany was in no way unique. In France 'many districts supported as many, or very nearly as many, inhabitants as at the beginning of the twentieth century'.[5] In the region of Oisans, south-east of Grenoble, there were about 13,000 inhabitants in 1339; by 1911 the total had risen only to 13,805. Around Neufbourg in the Eure a population of some 3,000 in 1310 was 3,347 as late as 1954. Around Elloe in the Fenland, settlement was almost as dense in 1260 as in 1951. In certain areas, in particular Artois, Flanders, Champagne and parts of Western Germany, the surplus population sought a solution to its problems in a move towards industrialisation. In the whole of Western Europe, villages grew into towns and cities with ten or twenty thousand inhabitants were no longer freakish rarities.[6] But the flow to the

towns drew off only a small part of the rising population in the
countryside.

So long as the growing population had unused land ready to
hand which could easily be exploited to produce more food, then
no unmanageable problems were posed. In certain areas—Basse-
Provence, Catalonia, Sweden and Scotland—this was still the case
until well on into the fourteenth century.[7] Europe, viewed as a
whole, still had a fair amount of under-developed territory even
as late as 1350. But in the great population centres, from which
the peasantry could not or would not move, the end of the thirteenth century was a period of acute crisis. The forests that remained were jealously conserved, the mountains offered no hope
to the would-be farmer. Productivity fell as erosion, lack of
manure, failure to let fields lie fallow or to rotate crops on
scientific principles, drained the goodness from the tired soil.
The population soared, more and more mouths had to be filled,
the gap between production and demand grew ever wider.

Taine's aphorism about the *Ancien Régime*: 'The people are like
men walking through a pond with water up to their mouths; at
the smallest depression of the ground or rise in the level of the
water, they will lose their footing, sink and drown' can be applied
as well to the peasant of the later Middle Ages. And in Europe of
the fourteenth century depressions of the ground seemed more
the rule than the exception. The climate played a major part in the
mischief seventy or eighty years before the Black Death. The
intense cold led to a striking advance of the glaciers, polar as well
as Alpine. High rainfall caused a rise in the level of the Caspian
Sea. The cultivation of cereals in Iceland and of the vine in England was crippled and virtually extinguished; wheat growing
areas were reduced in Denmark and the uplands of Provence.[8]

The most grave consequence was a series of disastrous harvests.
There were famines in England in 1272, 1277, 1283, 1292 and
1311.[9] Between 1315 and 1319 came a crescendo of calamity.
Almost every country in Europe lost virtually the whole of one
harvest, often of two or three. The lack of sun hindered the
production of salt by evaporation and thus made still more difficult the conservation of what meat there was. Even if there had
been food to store, facilities for storage did not exist. In England
wheat more than doubled in price. Cannibalism was a commonplace; the poor ate dogs, wrote one chronicler, cats, the dung of

doves, even their own children.[10] Ten percent of the population of Ypres died of starvation.[11] Nor was this the end: 1332 was another disastrous year for the crops and the period between 1345 and 1348 would have seemed uniquely unfortunate in any other century.

Before the Black Death, therefore, most of Europe was in recession or, at the very least, had ceased to advance. Colonisation stopped even where fresh fields lay open for the conquest. The *Drang nach Osten* petered out at the frontiers of Lithuania and Latvia. The cloth trade of Flanders and Brabant stagnated. The great fairs of the Champagne, indices of the economic health of a large and flourishing region, significantly declined.[12] The prices of agricultural produce were falling: agriculture was no longer the easy road to prosperity which it had been for the past two hundred years. Put in the simplest terms, Europe had outgrown its strength and was now suffering the physical and mental malaise which inevitably follows so intemperate a progress.

To what extent this recession was reflected in a drop in the population can only be guessed at. Famines on the scale which Europe had endured must at least have checked the hectic growth of the previous two centuries. The retreat from marginal lands which had already set in by 1320 or 1330 in Haute Provence, the Massif Central, Germany west of the Vistula and certain areas of England suggests that in these areas at least a decline must have begun long before the impact of the Black Death.[13] But there is little or no evidence of serious depopulation and no reason to doubt that the hungry mouths in almost every major population centre of Europe must still have been far too numerous for the exiguous supply of food. This disproportion was aggravated by the turmoil to which wars and civil disorders reduced great areas of France, Spain and Italy. The direct cost in human lives may not have been enormous but the destruction of crops and houses and the disruption of the life of the countryside seriously reduced production at a time when a larger food supply was as necessary as ever.

At the middle of the fourteenth century, therefore, chronic over-population was rendering intolerable the existence of many, if not a majority of Europeans. It is tempting to take a step further and see the Black Death as nature's answer to the problem of over-population, a Malthusian check to the over-exuberance of the preceding centuries. Reviewing a book by Georges Duby,[14] Professor Postan remarked that he had 'been especially gratified to read the passages in the book wherein the depression of the fourteenth century is represented as the consequence, perhaps even the nemesis, of the inordinate expansion of the preceding epoch'.[15] Viewed in this light, the Black Death is the nemesis that met a population which bred too fast for too long without first providing itself with the resources needed for such extravagance. Slicher Van Bath attributed the high death rate of the Black Death largely to the prolonged malnutrition which was the consequence of over-rapid growth.[16] If there had been no plague, the argument goes, then the population would, in the course of nature, have had to be reduced by other means.

But this line of reasoning should not be pushed too far. For one thing it is by no means universally accepted that medieval agriculture was incapable of supporting the population of the period. Certain authorities, indeed, claim that it could have fed many more without undue strain.[17] If there was no need in nature for the population to be reduced, then the Malthusian argument obviously falls to the ground. And even if it were accepted that Europe's population had outgrown its food supply by the middle of the fourteenth century it is still difficult to explain why the population should have continued to fall for a further fifty years or more. The check had worked, the hungry mouths were in the grave, even the most fanatic Malthusian would hardly have pleaded that the process should be continued.

Elizabeth Carpentier has summed up the controversy with her accustomed lucidity. 'Was the Black Death,' she asks, 'an evil made necessary by inescapable evolution? Or was it a tragic accident at variance with the normal advance of events?'[18] But to define a question satisfactorily is not necessarily to arrive at any answer; indeed, in medieval history, it sometimes seems that the more precisely a question is defined, the more certain it is that no answer will be forthcoming. Certainly in this case no clear-cut solution has been, or ever will be attained. All that can be said with confidence is that, in many parts of Europe, in the twelfth and thirteenth centuries, the population had grown with unusual speed; that this growth was a factor of importance, though by no means the only factor, which led to general malnutrition; that malnutrition was a contributary reason for the high death rate of the plague years; and that, as a result of the plague, the population was reduced to more easily manageable proportions. This humble conclusion leaves open many impassioning problems of what was cause and what effect; what blind chance and what the inexorable march of nature. It should be of comfort to future generations of historians to know that such problems exist and sobering for us to reflect that, even though we may triumphantly close the dossier with a decisive answer, our sons and grandsons will quickly have it open once again.

Whatever one's thesis about the inevitability of the Black Death it cannot be denied that it found awaiting it in Europe a population singularly ill-equipped to resist. Distracted by wars, weakened by malnutrition, exhausted by his struggle to win a living from his inadequate portion of ever less fertile land, the medieval peasant was ready to succumb even before the blow had fallen. But it was not only physically that he provided an easy prey; intellectually and emotionally he was prepared for disaster and ready to accept if not actually to welcome it.

Though the Europeans of the fourteenth century were painfully aware that they understood little of the disease which was destroying them they were at least confident that they knew the prime cause of their suffering. Few contemporary chroniclers fail to point out that the plague was an affliction laid on them by the Almighty, retribution for the wickedness of the present generation. Konrade of Megenberg, in his refreshingly heretical *Buch der Natur*,[19] was virtually unique in dismissing the theory of divine punishment on the grounds that nothing so promiscuous in its results could possibly have been intended by God.

It would have been astonishing if he had found many others to share his opinions. Even in the materialistic and, at a certain level, sophisticated nineteen-sixties the apocalyptic vision of a world about to incur destruction through its own folly and wickedness is by no means lost. Man-made devices may have been substituted for the pestilential hammer of the Middle Ages but both methods can be and are interpreted as manifestations of God's inscrutable workings.

How far more certain it was that the credulous and superstitious citizens of fourteenth-century Europe, unable to see any natural explanation of this sudden and horrifying holocaust, believing without question in hell-fire and the direct participation of God in life on earth, well-versed in Old Testament precedents for the destruction of cities or whole races in a sudden access of divine indignation, would take it for granted that they were now the victims of God's wrath. Like the citizens of Sodom and Gomorrah they were to die in expiation of their sins. 'Tell, O Sicily, and ye, the many islands of the sea, the judgments of God! Confess, O Genoa, what thou hast done, since we of Genoa and Venice are compelled to make God's chastisement manifest!'[20]

The Europeans were possessed by a conviction of their guilt. They were not so sure of what, exactly, they were guilty but the range of choice was wide. Lechery, avarice, the decadence of the church, the irreverence of the knightly classes, the greed of kings, the drunkenness of peasants; each vice was condemned according to the prejudices of the preacher and presented as the last straw which had broken the back of God's patience:

'In those days,' wrote the English chronicler, Knighton, 'there was much talk and indignation among the people because, when tournaments were held, in almost every place, a band of women would arrive as if they had come to join the sport, dressed in a variety of the most sumptuous male costumes. They used to wear partly-coloured tunics, one colour or pattern on the right side and another on the left, with short hoods and pendants like ropes wound round their necks, and belts thickly studded with gold and silver. They were even known to wear those knives which are called "daggers" in the vulgar tongue in pouches slung across their bodies; and thus they rode on choice war horses or other splendid steeds to the place of tournament. There and thus they spent or, rather, squandered their possessions, and wearied their bodies with fooleries and wanton buffoonery . . . But God, in this matter, as in all others, brought marvellous remedy . . .'[21]

Knighton was something of a conservative. Though many others might have condemned the current fashions, only a few would have called the Black Death a 'marvellous remedy' or have believed that God was being entirely temperate in his retribution when he obliterated quite so many to punish the extravagance of a few. But his conviction of the immorality of the age was widespread. 'These pestilences were for pure sin' wrote Langland sadly and more comprehensively.[22] None of those who believed that the plague was God's punishment of men suggested that the punishment did not fit the crime. God's will had to be done, his vengeance wreaked, and it was for man blindly to accept. To question His justice would have been a fresh and still more heinous sin, inviting yet further chastisement from on high.

Looking back, the victim of the Black Death saw a host of portents which should have warned him of God's intentions. Simon of Covino noticed heavy mists and clouds, falling stars, blasts of hot wind from the South. A column of fire stood above the papal palace at Avignon and a ball of fire was seen in the skies above Paris. In Venice a violent earth tremor set the bells of St Mark's pealing without touch of human hand. Anything which seemed in the least out of the way was retrospectively identified as a herald of the plague; a stranded whale, an outstandingly good crop of hazel nuts. Blood fell from bread when taken freshly from the oven. An illustration of the way that a legend could build up came in a later epidemic when mysterious bloodstains were found on men's clothes. Subsequent examination showed that the stains were in fact caused by the excrement of butterflies.

The skies not only provided a portent of what was coming but, through the movement of the planets, were the instruments by which the will of God was translated into harsh reality. In the fourteenth century astronomy was by far the most advanced branch of systematised scientific knowledge. For students of the stars, totally at a loss to explain what was happening around them, it was only natural to extrapolate desperately from what they understood and seek to compose from the movement of the planets some code of rules which would interpret and give warning of events on earth. 'The medieval cosmic outlook,' wrote Singer,[23] 'cannot be understood unless it is realized that analogy pushed to extreme lengths unchecked by observation and experiment was the major intellectual weapon of the age.'

Astrology, that arcane compound of astronomical research and semi-magical crystal-gazing, was near the peak of its prestige in the fourteenth century. It was the Arab astronomers who had evolved the theory that the movements of the planets and their relationship to each other in space dictated the future of humanity. Since the Black Death was clearly far out of the normal, some abnormal behaviour on the part of the planets had to be found to explain it.

Various theories were propounded from time to time but the classic exposition was that laid down by the Medical Faculty of the University of Paris in the report prepared on the orders of King Philip VI in 1348.[24] On 20 March, 1345, at 1pm, there occurred a conjunction of Saturn, Jupiter and Mars in the house of Aquarius. The conjunction of Saturn and Jupiter notoriously caused death and disaster while the conjunction of Mars and Jupiter spread pestilence in the air (Jupiter, being warm and humid, was calculated to draw up evil vapours from the earth and water which Mars, hot and dry, then kindled into infective fire). Obviously the conjunction of all three planets could only mean an epidemic of cataclysmic scale.

The doctrine that the movement of the planets was the force which set the Black Death in motion was never overtly challenged except by Konrade of Megenberg who argued that no planetary conjunction lasted for more than two years and that therefore, since the plague persisted longer, it must necessarily have had some other cause. Besides, he pointed out, all movements of celestial bodies were subject to strict order while the plague was patently haphazard in its action. Among a few other writers, however, a certain scepticism can be detected or, perhaps more correctly, an indifference to remote causes which were not susceptible of proof and were anyhow beyond the power of man to mend. Gentile da Foligno referred to the planets in general terms and then went on[25] '. . . It must be believed that, whatever may be the case with regard to the aforesaid causes, the immediate and particular cause is a certain poisonous material which is generated about the heart and lungs.' The job of the doctor, he concluded, was not to worry about the heavens but to concentrate on the symptoms of the sick and to do what he could to cure them.

Such admirable common sense was the exception. The European, in the face of the Black Death, was in general overwhelmed by a sense of inevitable doom. If the plague was decreed by God and the inexorable movement of the planets, then how could frail man seek to oppose it? The preacher might counsel hope, but only with the proviso that the sins of man must first be washed away by the immensity of his suffering. The doctor might prescribe remedies, but with the tepid enthusiasm of a civil-defence expert advising those threatened by imminent nuclear attack to adopt a crouching posture and clasp their hands behind their necks. The Black Death descended on a people who were drilled by their theological and their scientific training into a reaction of apathy and fatalistic resignation. Nothing could have provided more promising material on which a plague might feed.

THREE

Italy

The Black Death arrived in Sicily early in October, 1347, about three months before it reached the mainland. According to Michael of Piazza,[1] a Franciscan friar who wrote his history some ten years later, twelve Genoese galleys brought the infection to the port of Messina. Where they came from is unknown; possibly also from the Crimea, though they must have left the area several months before the galleys which bore the Black Death from Caffa to Genoa and Venice. Nor can one now know whether the disease was borne by rats and fleas or was already rampant among members of the crew; the chronicler's description of 'sickness clinging to their very bones' suggests the latter.

Within a few days the plague had taken a firm grasp on the city. Too late to save themselves, the citizens turned on the sailors who had brought them this disastrous cargo and drove them from the port. With their going, the Black Death was scattered around the Mediterranean but Messina's sufferings were no lighter for its dispersion. With hundreds of victims dying every day and the slightest contact with the sick seeming a guarantee of rapid infection, the population panicked. The few officials who might have organised some sort of measures to mitigate the danger were themselves among the first to perish. The people of Messina fled from their doomed city into the fields and vineyards of southern Sicily, seeking safety in isolation and carrying the plague with them through the countryside.

When the first victims reached the neighbouring city of Catania they were lodged in the hospital and kindly treated. But as soon as the Catanians realised the scale and nature of the disaster they concluded that, by accepting the refugees, they were condemning themselves to the same fate. Strict control over immigration was

introduced and it was decreed that any plague victim who had
already arrived and subsequently perished should be buried in
pits outside the walls. 'So wicked and timid were the Catanians,'
wrote Michael of Piazza, 'that they refused even to speak to any
from Messina, or to have anything to do with them, but quickly
fled at their approach.' What was wickedness in the eyes of the
doomed of Messina must have seemed elementary prudence to
the menaced of Catania. The same pattern of behaviour was to
be repeated all over Europe but rarely did it do any good to those
who sought to save themselves by cutting themselves off from
their neighbours. The Black Death had already breached the
walls of Catania and nothing could stop it running riot through
the population.

The Messinese now appealed to the Patriarch Archbishop of
Catania to allow the relics of St Agatha to be taken from Catania
to Messina. The Patriarch agreed but the Catanians, not un-
naturally feeling that charity began at home and that St Agatha
should remain at her post in her own cathedral, rose in protest.
'They tore the keys from the sacristan and stoutly rebuked the
Patriarch, saying that they would rather die than allow the relics
to be taken to Messina.' The Patriarch, who must have been a man
of singular courage, accepted the mob's decision but insisted at
least on dipping some of the relics in water and personally taking
the water with him to Messina.

'The aforesaid Patriarch,' reads Michael of Piazza's account,
'landed at Messina carrying with him the holy water . . . and in
that city there appeared demons transfigured into the shape of
dogs, who wrought grievous harm upon the bodies of the
citizens; so that men were aghast and dared not go forth from
their houses. Yet by common consent, and at the wish of the
Archbishop, they determined to march devoutly around the city
reciting litanies. While the whole population was thus processing
around the streets, a black dog, bearing a drawn sword in his
paws, appeared among them, gnashing with his teeth and rushing
upon them and breaking all the silver vessels and lamps and
candlesticks on the altars, and casting them hither and thither . . .
So the people of Messina, terrified by this prodigious vision, were
all strangely overcome by fear.'

This description exemplifies the curious blend of sober eye-
witness reporting and superstitious fantasy which is characteristic

of so many similar chronicles. How much of it should one believe? How much of it, for that matter, did Michael of Piazza
believe himself? Rabies was endemic in Sicily and, since nobody
can have had the time or energy to keep mad dogs in check, it is
not surprising that there should have been an unusually large
number running in the streets. In the circumstances the panic-
stricken Sicilian can hardly be blamed for detecting some supernatural influence in their activities. But did the chronicler really
believe that he, or some other reliable witness had actually seen
a black dog with a drawn sword in its paws? Or was the statement no more than an expression in symbolic terms of the
chronicler's belief in the dog's daemonic possession? Probably
Michael of Piazza himself would hardly have known the answer.
Medieval man skated on the thinnest possible ice of verified
knowledge with beneath him unplumbed and altogether terrifying depths of ignorance and superstition. Let the ice break and
with it was lost all grasp on reality and all capacity for objective,
logical analysis.

The people of Messina, disappointed of St Agatha's relics, then
set off barefoot in procession to a shrine some six miles away
where was to be found an image of the Virgin said to possess
exceptional powers. Once again they were discomfited and, in
their discomfiture, Michael of Piazza saw another proof that the
plague was God's retribution on his erring people.

'This aforesaid Mother of God, when she saw and drew near
unto the city, judged it to be so hateful and so profoundly
stained with blood and sin that she turned her back upon it, being
not only unwilling to enter therein, but even abhorring the very
sight thereof. For which cause the earth yawned open and the
horse which bore the image of the Mother of God stood fixed
and motionless as a rock.'

Eventually the animal was bullied or cajoled into the city and
the Virgin lodged in Santa Maria la Nuova, the largest church of
the city. But little good did it do the unfortunate Messinese.
'. . . this coming of the image availed naught; nay, the pestilence
raged so much the more violently that one man could not succour
another but the greater part of the citizens deserted Messina and
were scattered abroad.'

After his return from Messina the gallant Patriarch succumbed to the disease which he had combated so stoutly. He was buried in the Cathedral at Catania. By his behaviour and by his death he set a standard which was to be matched by few indeed of his peers.

Quickly the plague spread over Sicily, ravaging with especial violence the towns and villages at the western end. It was not for long confined to such narrow limits. Sicily, as Professor Renouard dryly remarks,[2] 'fulfilled its natural mission as a centre of the Mediterranean world'. From thence it spread probably to North Africa by way of Tunis; certainly to Corsica and Sardinia; the Balearics, Almeria, Valencia and Barcelona on the Iberian peninsula; and to Southern Italy. It is remarkable, in this as in every other epidemic of bubonic plague, how closely the disease followed the main trade-routes.[3] Largely, of course, this is a token of the role which the rat played in the propagation of plague. But whether the Black Death travelled by rat, by unescorted flea or by infected sailor, ship was the surest and most rapid means. The Black Death, indeed, is peculiar among plagues in that the particularly high incidence of its pneumonic variant meant that it struck inland with unusual vigour. But even though it could thus attain the hinterland, its first target was still the coastal towns. It travelled from the Crimea to Moscow not overland but by way of Italy, France, England and the Hanseatic ports.[4]

The three great centres for the propagation of the plague in Southern Europe were Sicily, Genoa and Venice. It seems to have arrived more or less simultaneously at the latter ports some time in January, 1348. But it was Pisa, attacked a few weeks later,[5] which provided the main point of entry to Central and Northern Italy. From there it moved rapidly inland to Rome and Tuscany. It had begun the march which was not to end until the whole continent of Europe had been blanketed by death.

In Italy the previous years had provided a chapter of disasters less dramatic but little less damaging than those which had overtaken the unfortunate Chinese. A crescendo of calamity was reached shortly before the plague arrived.[6] Earthquakes had done severe damage in Naples, Rome, Pisa, Bologna, Padua and Venice. The wine in the casks had become turbid: 'A statement which,' as the nineteenth-century German historian Hecker hopefully remarked, 'may be considered as furnishing a proof that changes causing a decomposition of the atmosphere had taken place.' From July, 1345, six months of almost continuous rain had made sowing impossible in many areas. The following spring things were little better. The corn crop was less than a quarter of the usual and almost all the domestic fowls had to be slaughtered for want of feeding stuffs. Even for the richest states and cities it was difficult to replace the loss by imports. 'In 1346 and 1347 there was a severe shortage of basic foodstuffs . . . to the point where many people died of hunger and people ate grass and weeds as if they had been wheat.'[7] Near Orvieto the bridges were washed away by the floods and the damage done to communications all over Italy made the work of feeding the hungry still more difficult.[8]

Inevitably prices soared. The price of wheat doubled in the six months prior to May, 1347, and even bran became too costly for the poor. In April, 1347, a daily ration of bread was being issued to 94,000 people in Florence; prosecutions for all minor debts were suspended by the authorities and the gates of the prisons thrown open to all except serious criminals. It is said that four thousand Florentines died either of malnutrition or from diseases which, if malnutrition had not first existed, would never have proved fatal.[9] And yet of all the cities of Italy, Florence, with its great wealth, its powerful and sophisticated administration and its relatively high standards of education and of hygiene, was best equipped to cope with the problems of famine and disease.

Financial difficulties in Florence and Siena, which the agricultural problem complicated but did not create, made things even worse. The great finance house of the Peruzzi was declared bankrupt in 1343, the Acciaiuoli and the Bardi followed in 1345. By 1346 the Florentine houses alone had lost 1.7 million florins and virtually every bank and merchant company was in difficulties. It was an economic disaster without precedent.[10] Even if the grain had been available it would have been hard for the cities of Tuscany to find the money to purchase it.

The final and perhaps the most dangerous element in this sombre picture was the political disorder which was an almost invariable feature of fourteenth-century Italy. There were, said

Professor Caggese,[11] no 'events of universal import' but only a multiplicity of 'local dramas.' These dramas turned Italy into a bloody patch-work of bitter and seemingly unending squabbles. The Guelphs fought the Ghibellines, the Orsini fought the Colonna, Genoa fought Venice, the Visconti fought everybody and marauding German freebooters preyed on what was left. Rome was demoralised by the disappearance of the Papacy to Avignon and shaken by the revolution of Rienzo. Florence had recently experienced the rising of Brandini. Naples was in turmoil as Lewis of Hungary pursued his vendetta against Queen Joanna, the murderer of his brother.

For the nobles and the warriors there was, at least, glamour, excitement and a chance of booty. For the common people there was nothing except despairing fear, a total and disastrous lack of confidence in what the future might hold for them. What has been argued of Europe as a whole is, *a fortiori*, true of Italy. The people were physically in no state to resist a sudden and severe epidemic and psychologically they were attuned to an expectation and supine acceptance of disaster. They lacked the will to fight; almost, one might think, they welcomed the termination of their troubles. To speak of a collective death-wish is to trespass into the world of metaphysics. But if ever there was a people with a right to despair of life, it was the Italian peasantry of the midfourteenth century.

'Oh, happy posterity,' wrote Petrarch of the Black Death in Florence, 'who will not experience such abysmal woe and will look upon our testimony as a fable.'[12] The Black Death is associated more closely with Florence than with any other city; so much so that in contemporary and even more recent accounts it is sometimes referred to as 'The Plague of Florence.' Partly this is because Florence at that period was one of the greatest cities of Europe and certainly the first of them to feel the full force of the epidemic. Partly it is because the plague raged there with exceptional intensity; certainly more severely than in Rome, Paris or Milan and at least as violently as in London or Vienna. But most of all Florence owes its notoriety to the terms in which its sufferings were described. In his introduction to *The Decameron* Boccaccio wrote what is undoubtedly and deservedly the best-

known account of the Black Death and probably the most
celebrated eye-witness account of any pestilence in any epoch.[13]
One or two sentences from it have already appeared in this book
but no account of the Black Death would be complete unless it
were quoted extensively.

'In Florence,' wrote Boccaccio, 'despite all that human wisdom
and forethought could devise to avert it, as the cleansing of the
city from many impurities by officials appointed for the purpose,
the refusal of entrance to all sick folk, and the adoption of many
precautions for the preservation of health; despite also humble
supplications addressed to God, and often repeated both in
public procession and otherwise, by the devout; towards the
beginning of the spring of the said year the doleful effects of the
pestilence began to be horribly apparent by symptoms that shewed
as if miraculous.

'. . . Which maladies seemed to set entirely at naught both the
art of the physician and the virtues of the physic; indeed, whether
it was that the disorder was of a nature to defy such treatment, or
that the physicians were at fault—besides the qualified there was
now a multitude both of men and women who practised without
having received the slightest tincture of medical science—and,
being in ignorance of its source, failed to apply the proper
remedies; in either case . . . almost all . . . died, and in most cases
without any fever or other attendant malady . . .

'In which circumstances . . . divers apprehensions and imagina-
tions were engendered in the minds of such as were left alive;
inclining almost all of them to the same harsh resolution; to wit,
to shun and abhor all contact with the sick and all that belonged
to them, thinking thereby to make each his own health secure.
Among whom there were those who thought that to live temper-
ately and avoid all excess would count for much as a preservative
against seizures of this kind. Wherefore, they banded together,
and, disassociating themselves from all others, formed communities in houses where there were no sick, and lived a separate
and secluded life, which they regulated with the utmost care,
avoiding every kind of luxury, but eating and drinking very
moderately of the most delicate viands and the finest wines,
holding converse with none but one another, lest tidings of sickness or death should reach them, and diverting their minds with

music and such other delights as they could devise. Others, the bias of whose minds was in the opposite direction, maintained that to drink freely, to frequent places of public resort, and to take their pleasure with song and revel, sparing to satisfy no appetite, and to laugh and mock at no event, was the sovereign remedy for so great an evil: and that which they affirmed they also put into practice, so far as they were able, resorting day and night now to this tavern, now to that, drinking with an entire disregard of rule or measure, and by preference making the houses of others, as it were, their inns, if they but saw in them aught that was particularly to their taste or liking; which they were readily able to do because the owners, seeing death imminent, had become as reckless of their property as of their lives; so that most of the houses were open to all comers, and no distinction was observed between the stranger who presented himself and the rightful lord . . . In this extremity of our city's sufferings and tribulation the venerable authority of laws, human and divine, was abused and all but totally dissolved, for lack of those who should have administered and enforced them, most of whom, like the rest of the citizens, were either dead or sick or so hard beset for servants that they were unable to execute any office; whereby every man was free to do what was right in his own eyes.

'Not a few there were who belonged to neither of the two said parties, but kept a middle course between them . . . living with a degree of freedom sufficient to satisfy their appetites, and not as recluses. They therefore walked abroad, carrying in their hands flowers or fragrant herbs or divers sorts of spices, which they frequently raised to their noses, deeming it an excellent thing thus to comfort the brain with such perfumes, because the air seemed to be everywhere laden and reeking with the stench emitted by the dead and dying, and the odours of drugs.

'Some again, the most sound, perhaps, in judgement, as they were also the most harsh in temper, affirmed that there was no medicine for the disease superior or equal in efficiency to flight; following which prescription a multitude of men and women, negligent of all but themselves, deserted their city, their houses, their estates, their kinsfolk, their goods, and went into voluntary exile, or migrated to the country, as if God, in visiting men with this pestilence in requital of their iniquities, would not pursue them with His wrath wherever they might be, but intended the destruction of such alone as remained within the circuit of the walls of the city . . .

'. . . Tedious were it to recount how citizen avoided citizen, how among neighbours was scarce found any that showed fellow-feeling for another, how kinsfolk held aloof and never met, or but rarely; enough that this sore affliction entered so deep into the minds of men and women that, in the horror thereof, brother was forsaken by brother, nephew by uncle, brother by sister and, oftentimes, husband by wife; nay, what is more and scarcely to be believed, fathers and mothers were found to abandon their own children, untended, unvisited, to their fate, as if they had been strangers . . . In consequence of which dearth of servants and dereliction of the sick by neighbours, kinsfolk and friends, it came to pass—a thing, perhaps, never before heard of—that no woman, however dainty, fair or well-born, shrank, when stricken by the disease, from the ministrations of a man, no matter whether he were young or no, or scrupled to expose to him every part of her body, with no more shame than if he had been a woman, sub-mitting of necessity to that which her malady required; where-from, perchance, there resulted in after-time some loss of mod-esty in such as recovered . . .

'It had been, as to-day it still is, the custom for the women that were neighbours or of kin to the deceased to gather in his house with the women that were most closely connected with him, to wail with them in common, while on the other hand his male kinsfolk and neighbours . . . assembled without, in front of the house, to receive the corpse; and so the dead man was borne on the shoulders of his peers, with funeral pomp of taper and dirge, to the church selected by him before his death. Which rites, as the pestilence waxed in fury, were either in whole or in great part disused and gave way to others of a novel order. For not only did no crowd of women surround the bed of the dying, but many passed from this life unregarded, and few indeed were they to whom were accorded the lamentations and bitter tears of sorrow-ing relations; nay, for the most part, their place was taken by the laugh, the jest, the festal gathering; observances which the women, domestic piety in large measure set aside, had adopted with very great advantage to their health. Few also there were whose bodies were attended to the church by more than ten or twelve neighbours, and those not the honourable and respected citizens, but a sort of corpse-carrier drawn from the baser ranks, who called themselves *becchini* and performed such offices for hire, would shoulder the bier and, with hurried steps, carry it, not to the church of the dead man's choice, but to that which was nearest at hand, with four or six priests in front and a candle or two, or, perhaps none; nor did the priests distress themselves with too long and solemn an office, but with the aid of the *becchini* hastily consigned the corpse to the first tomb which they found untenanted . . . Many died daily or nightly in the public streets; of many others, who died at home, the departure was hardly observed by their neighbours, until the stench of their putrefying bodies carried the tidings; and what with their corpses and the corpses of others who died on every hand the whole place was a sepulchre.

'It was the common practice of most of the neighbours, moved no less by fear of contamination by the putrefying bodies than by charity towards the deceased, to drag the corpses out of the houses with their own hands, aided, perhaps, by a porter, if a porter was to be had, and to lay them round in front of the doors, where any one that made the round might have seen, especially in the morning, more of them than he could count; afterwards they would have biers brought up or, in default, planks whereon they laid them. Nor was it only once or twice that one and the same bier carried two or three corpses at once; but quite a considerable number of such cases occurred, one bier sufficing for husband and wife, two or three brothers, father and son, and so forth. And times without number it happened that, as two priests bearing the cross were on their way to perform the last office for some one, three or four biers were brought up by the porters in rear of them, so that, whereas the priests supposed that they had but one corpse to bury, they discovered that there were six, or eight, or sometimes more. Nor, for all their number, were their obsequies honoured by either tears, or lights, or crowds of mourners; rather it was to come to this, that a dead man was then of no more account than a dead goat would be to-day . . .

'As consecrated ground there was not in extent sufficient to provide tombs for the vast multitude of corpses which day and

night, and almost every hour, were brought in eager haste to the
churches for interment, least of all, if ancient custom were to be
observed and a separate resting-place assigned to each, they dug
for each graveyard, as soon as it was full, a huge trench in which
they laid the corpses as they arrived by hundreds at a time, piling
them up as merchandise is stowed in the hold of a ship, tier upon
tier, each covered with a little earth, until the trench would hold
no more. But I spare to rehearse with minute particularity each
of the woes that came upon our city, and say in brief that, harsh
as was the tenor of her fortunes, the surrounding country knew
no mitigation; for there—not to speak of the castles, each, as it
were, a little city in itself—in sequestered villages, or on the open
champaign, by the wayside, on the farm, in the homestead; the
poor, hapless husbandmen and their families, forlorn of physician's care or servants' tendance, perished day and night alike,
not as men but rather as beasts. Wherefore they too, like the
citizens, abandoned all rule of life, all habit of industry, all
counsel of prudence; nay, one and all, as if expecting each day to
be their last, not merely ceased to aid Nature to yield her fruit in
due season of their beasts and their lands and their past labours,
but left no means unused, which ingenuity could devise, to waste
their accumulated store; denying shelter to their oxen, asses,
sheep, goats, pigs, fowls, nay even to their dogs, man's most faithful companions, and driving them out into the fields to roam at
large amid the unsheaved, nay unreaped corn . . .

'But enough of the country! What need we add but (reverting
to the city) that . . . it is believed without any manner of doubt,
between March and the ensuing July, upwards of a hundred
thousand human beings lost their lives within the walls of the
city of Florence, which before the deadly visitation would not
have been supposed to contain so many people! How many grand
palaces, how many stately homes, how many splendid residences,
once full of retainers, of lords, of ladies, were now left desolate
of all, even to the meanest servant! How many families of historic
fame, of vast ancestral domains and wealth proverbial, found now
no scion to continue the succession! How many brave men, how
many fair ladies, how many gallant youths, whom any physician,
were he Galen, Hippocrates or Aesculapius himself, would have
pronounced in the soundest of health, broke fast with their kins-

folk, comrades and friends in the morning, and when evening came, supped with their forefathers in the other world!'

Boccaccio used this description as the preamble to his *Decameron*; a stark background against which he was to create a miracle of light and vivid fantasy. It is only reasonable to consider whether, in the interests of dramatic contrast, he did not portray the Black Death in Florence in even gloomier colours than it deserved. Certainly he was not anxious to stress the happier side: the selfless devotion of certain nuns or doctors, the efforts of the city government to keep going some sort of order and administration. Certainly, too, few cities suffered as much as Florence. But so much of Boccaccio's detail is to be found in the records of contemporary chroniclers in France, Germany and England as well as Italy, that no one can doubt its essential truth.

The headlong flight from the cities, abandoning possessions and leaving houses open to all the world; the ruthless desertion of the sick, to meet their end as best they might, with no company but their own; the hurried, sordid burials in great communal pits; crops wasting in the fields and cattle wandering untended over the countryside—such details are the common currency of the chroniclers. On some points, even, it seems that Boccaccio does not do full justice to the horror: other reports, for instance, give more attention to the sinister role of the *becchini*,[14] brutalised monsters, their life not worth twenty-four hours' purchase, who would force their way into the houses of the living and tear them away to join the dead unless the men paid for their safety with a handsome bribe or the women with their virtue.

In its picturesque detail, therefore, one must accept Boccaccio's account as accurate and authentic. But the same cannot be said for his statistics. His estimate of a hundred thousand dead within the city is patently exaggerated. By 1345 the population of Florence was already declining from its zenith of some fifty years before. The evidence of the number of bread tickets issued in April, 1347, suggests a population of well over ninety thousand[15] and the most authoritative modern estimate similarly puts it at between eighty-five and ninety-five thousand, with a slight preference for the higher figure.[16] Unless Florence was virtually unique it seems impossible that more than two thirds and un-

likely that much more than half of these can have died during the
six months of the plague. In the much smaller but in many ways
comparable cities of San Gimignano, Siena and Orvieto, analysis
of the available data suggests a death rate of about 58% in the
first[17] and 50% (or a little more) in the others.[18] One could not be
far wrong if one guessed that between forty-five and sixty-five
thousand Florentines died of the Black Death.

Boccaccio's estimate, though extravagant, was not wholly fantastical. It is noteworthy, too, that he qualified it with some surprise that the population of the city should have turned out to be
so much greater than had been generally believed. In this he was
more cautious than many of his contemporaries who manipulated
or invented statistics with almost inconceivable levity. Dr
Coulton has referred to the 'chronic and intentional vagueness'
of the medieval mind when confronted by a set of figures and
quotes as an example the action of the English parliament which,
in 1371, fixed the level of a tax on the basis that there were some
forty thousand parishes in the kingdom, while, in fact, the most
cursory study of readily available records would have shown that
there were less than nine thousand.[19]

Partly this may have been due to the intractability of Roman
numerals for complicated multiplication or division but there
seems too to have been genuine indifference to the need for, indeed the very possibility of, precision. A large figure was a picturesque adornment to an argument but not part of the basic data
from which a conclusion was drawn. It might be expressed as
though exactly calculated but this was merely so as to heighten
the dramatic effect. When the Pope was assured by his advisers
that the Black Death had cost the lives of 42,836,486 people
throughout the world, or the losses in Germany were estimated at
1,244,434,[20] what was meant was that an awful lot of people had
died.

The estimates of chroniclers are not always so nonsensical.
When the Chronicler of Este[21] said that, in and around Naples,
sixty-three thousand people were killed by the plague in two
months, the figure was high but not impossible. It is again unlikely that the Chronicler of Bologna was right in saying that three
out of every five people died[22] but there are contemporary historians who maintain that in certain Italian cities the mortality
rate was in the region of 60%.[23] But, right or wrong, neither of

these writers was convinced of or even particularly concerned about the literal accuracy of his figures; the estimate was an expression in vivid and easily remembered form of the enormity of his experience. The material for an even slightly accurate census did not exist and the contemporary scholar, extrapolating from a few verified facts, is more likely to arrive at a sensible answer than the medieval chronicler dependent on his own eyes and a vivid imagination and convinced, anyway, that the matter was one of trivial importance.

Though the Florentines were subjected to almost intolerable pressure it does not seem that the machinery of government ever broke down altogether. The same is true of the other Italian cities. Venice was one of the first to be afflicted; not surprisingly, since its position as chief European port of entry for goods from the East was bought at the cost of seventy major epidemics in seven hundred years.

At the worst of the Black Death six hundred Venetians a day were said to be dying; this rate can hardly have been sustained for long but is by no means incredible.[24] On 20 March, 1348, the Doge, Andrea Dandolo, and the Great Council appointed a panel of three noblemen to consider measures to check the spread of the plague.[25] A few days later the panel reported its recommendations. Remote burial places were designated; one at S Erasmo at what is now the Lido, another on an island called S Marco Boccacalme which seems since to have vanished into the lagoon.[26] A special service of barges was provided to carry corpses to the new graveyards. All the dead, it was ruled, were to be buried at least five feet underground. Within the city itself beggars were forbidden to exhibit corpses in the streets, as was their macabre custom, and various measures of relief were adopted, including the release of all debtors from gaol. Surgeons were exceptionally allowed to practise medicine. Strict control of immigration was introduced and any ship which tried to evade it was threatened with burning. Either in this epidemic or possibly the next a quarantine station was set up at the Nazarethum where voyagers returning from the Orient were isolated for forty days—the period, apparently, being fixed by analogy to Christ's suffering in the wilderness.

But such precautions, even though the Great Council did what it could to enforce them, came too late to save the city. The dead, as in Florence, were numbered at a hundred thousand; the estimate seems to have had even less in the way of a factual basis. Doctors, in particular, suffered and within a few weeks almost all were dead or had fled the city. A certain Francesco of Rome was a Health Officer in Venice for seventeen years. When he retired he received an annuity of twenty-five gold ducats as a reward for staying in Venice during the Black Death 'when nearly all physicians withdrew on account of fear and terror.' When asked why he did not flee with the rest he answered proudly: 'I would rather die here than live elsewhere.'[27]

Even harsher measures of control were tried in other cities. In Milan, when cases of the plague were first discovered, all the occupants of the three houses concerned, dead or alive, sick or well, were walled up inside and left to perish.[28] It is hard to believe that this drastic device in fact served any useful purpose but for this or some other reason the outbreak seems to have been postponed by several weeks and Milan was the least afflicted of the large Italian cities. The technique of entombment was sometimes employed by the householder as well as by the authorities. In Salé, Ibn al Khatīb records, Ibn Abu Madyan walled up himself, his household and a plentiful supply of food and drink and refused to leave the house until the plague had passed. His measures were entirely successful: a fact disturbing to those who believed that the atmosphere had been corrupted and that bricks and mortar could consequently be no impediment to the disease.

Boccaccio, in fact, did less than justice to the efforts of the city fathers to control the plague in Florence. A committee of eight was set up from among the wisest and most respected citizens and was given something close to dictatorial powers. But when it came to the point there was not much which even the wisest of committees could achieve. Their regulations were largely concerned with the removal of decaying matter from the markets and the dead and dying from the streets—sensible enough objects but, by themselves, not likely to save the city. Nor, when the plague was at its worst, did the personnel exist to ensure that even so modest a policy was carried out.

Pistoia, where the civic ordinances published during the Black

Death have been preserved, provides an unusually clear picture both of the efforts which the authorities made to preserve the citizens and the limitations imposed by their ignorance and the weakness of the machinery of government.[29] On 2 May, 1348, when the first cases of the Black Death were beginning to appear in the vicinity, the Council enacted nine pages of regulations intended to guard the town against infection. No one was to visit the Pisan or Luccan states where the plague was already rampant. If such a visit were made then, even though the citizen had started his journey before the date of the ordinances, his return to Pistoia was forbidden. No linen, woollen goods, or, not surprisingly, corpses, whatever their source, were to be imported into the town. Food markets were put under strict supervision. Attendance at funerals was to be limited to members of the family and standards were laid down for the place and depth of burials. To avoid disturbing the sick and also, no doubt, so as not to undermine the morale of the healthy, there was to be no tolling of bells at funerals and no announcements by criers or trumpeters. The second set of ordinances, of 23 May, relaxed the ban on travel; no doubt because the plague had now taken so firm a grip that any such precautions would have been futile. Rules for the supervision of markets, however, were further tightened up. On 4 June changes were made to the rules for funerals. Sixteen men from each part of the town were to be selected as grave diggers and nobody else was to be allowed to do such work. Because of the shortage of wax, candles were no longer to be burned for the dead. Finally, on 13 June, the regulations governing the defence of the city were re-cast. So as to spare the cavalry, who were traditionally drawn from the richer section of society, 'the fatigue of mind and body which had been proved to induce pestilence,' it was decreed that each cavalry man could provide a substitute to perform his duties.

This last proviso is of interest as being one of the very few instances of legislation or any other kind of official pronouncement which discriminated in favour of the rich and noble. Obviously the rich were better equipped than the poor to protect themselves against the plague but the temptation to Church and State to load the dice still further in their favour was generally resisted. Indeed, on the whole, civic authorities and national governments alike seem to have accepted their responsibilities towards the poorer sections of their populations and to have done their inadequate best to shield them from disaster.

A contrast to the responsible attitude of the Pistoians is to be found in the apparent apathy of the government of Orvieto. In her profound and brilliant study of Orvieto at the time of the Black Death,[30] Dr Elizabeth Carpentier has analysed the impact of the disease and the reactions of its victims in terms which, *mutatis mutandis*, must be valid for every medium-sized town of Italy.

Orvieto in the mid-fourteenth century was a small but prosperous town of about 12,000 inhabitants. It had lost even more than its neighbours in the perpetual warfare between Guelph and Ghibelline and its rich vineyards and wheat fields had repeatedly been ravaged by marauding visitors or discontented citizens. The insecurity from which the whole region suffered had done serious harm to its role as a commercial centre and had reduced to almost nothing the profit which it gained from transit trade. The famine of 1346 and 1347 hit Orvieto badly though the hardships endured by its citizens were small compared with those in other, less well provided regions. By the autumn of 1347 the worst seemed to be over. There had been a good harvest and what promised to be a reasonably stable peace had been patched up. But the situation was still precarious and the powers of resistance of the average Orvietan had been gravely sapped.

Then, in the spring of 1348, came the Black Death. For months previously the members of the Council must have known that disaster was on the way but no official discussion took place and no preventive measures were adopted. When the Council met on 12 March, 1348, the plague had reached Florence, some eighty miles away. But still the official silence was maintained. Perhaps the Councillors believed, not without reason, that it did not lie in their power to avert disaster and that, therefore, the less said the better. But there was also something of the spirit of the child who sees moving up on him the thundercloud that will ruin his play but says nothing in the hope that, if he pretends not to notice it, it will miraculously move away.

The Council, in fact, had good reason to think that there was little or nothing for them to do. Orvieto had one doctor and one surgeon paid by the city to tend the poor and teach students. There were seven private doctors listed as owning land and perhaps two or three more besides. This was not a bad tally for a town of medium size. But the hospitals were less impressive. Only one was reasonably well financed; the others had little space and less facilities. And the state of public hygiene was deplorable. Constantly reiterated laws against rearing pigs and goats in the street, tanning skins in mid-city and throwing refuse out of windows show that the Council was concerned about the situation but also that it was powerless to improve it. Dirt and malnutrition were the two great allies of the plague, in Orvieto as in so many other cities.

The Black Death was probably brought to Orvieto in the train of the Ambassador who arrived from Perugia towards the end of April, 1348. It raged for four months at something near to full strength and reached its peak in July. It seems that the septicaemic form of the disease was common since there are many references to people dying within twenty-four hours of the first symptoms appearing. According to a contemporary chronicler more than five hundred victims died a day at the worst of the mortality and the final death roll included more than ninety percent of the population. Both figures are unacceptably high; the first particularly so since even a week of such intensity would have eliminated more than a quarter of the total population. Dr Carpentier, working from admittedly scanty material, is inclined to put the mortality rate at about fifty percent.

Compared with ninety percent such a figure may seem tolerable, but it is still difficult to conceive the impact of a catastrophe which, within four or five months, removed every second person from a small and closely knit community. In every family of four in Orvieto, one of the parents and one of the children could statistically expect to die. The surprise is not that there was panic and despair but that the fabric of the city's social life survived more or less unimpaired. Until the end of June the official records made no mention of the plague. Business as usual was the order of the day. Then the strain became too great. Of the Council of Seven elected at the end of June, two were dead by 23 July, three more by 7 August. Another member was ill by 10 August and, though he seems eventually to have recovered, a further death was recorded by 21 August. From the beginning of July all regular Council meetings were abandoned and there were blank pages in the city register. In mid-August Orvieto's most important religious ceremony, the procession of the Assumption, was abandoned on orders of the authorities. Yet by the end of the month the Council was meeting regularly once more, the shops were open, three new public notaries and two new gate-keepers had been appointed, the many problems arising out of wills and intestacies were being sorted out. For the doctors, at least, there was a silver lining to the cloud. Before the Black Death the physician retained by the city, if he was lucky enough even to receive his salary, was paid £25 a year. After the plague Matteo fù Angelo was offered £200 a year and exemption from all the civic taxes.

Orvieto survived; to the outward eye at least substantially un-scathed. Siena was left with a visible memento of the plague for posterity to wonder at. In 1347 work was in progress on what was to be the greatest church of Christendom. The transept of the Cathedral was built, the foundations of the choir and nave laid out. Then came the Black Death. The workmen perished, the money was diverted to other more urgent purposes. When the epidemic passed the shattered city could not find the funds or energy to complete the project. The truncated body of the Cathedral remained, was patched up and gradually became so much an accepted part of the landscape that to-day it is hard to believe it was ever intended to take a different form.

'Father abandoned child;' wrote Agnolo di Tura[31] of the plague at Siena, 'wife, husband; one brother, another; for this illness seemed to strike through the breath and the sight. And so they died. And no one could be found to bury the dead for money or for friendship . . . And in many places in Siena great pits were dug and piled deep with huge heaps of the dead . . . And I, Agnolo di Tura, called the Fat, buried my five children with my own hands, and so did many others likewise. And there were also many dead throughout the city who were so sparsely covered with earth that the dogs dragged them forth and devoured their bodies.'

According to Agnolo di Tura, if his somewhat convoluted cal-culations have been interpreted correctly, fifty thousand died within the city including thirty-six thousand old people. Many more fled to the country and, when the Black Death passed on, only ten thousand inhabitants remained. Since the total population of Siena could not, at the most, have exceeded fifty thousand in

1348 one finds, once again, a contemporary estimate which is not only improbable but actually impossible. But there is plenty of evidence that the city was unusually hard hit. The wool industry was closed down and the import of oil suspended. On 2 June, 1348, all civil courts were recessed by the City Council, not to re-open till three months later. In an emergency session of the Council, legalised gambling was prohibited 'forever'; the loss of revenue was considerable and, as it turned out, eternity was deemed to have run its course before the end of the year. The size of the City Council was reduced by a third and the obligatory quorum of members cut to a half. The church waxed fat from inheritances and gifts from frightened citizens; so much so that, in October, all annual appropriations to religious persons and institutions were suspended for two years.

Siena is an example of a city which, superficially, recovered quickly from the Black Death but, in reality, suffered economic and political dislocation so profound that things were never to be the same again. An intensive campaign to attract immigrants by tax concessions and other devices filled many of the gaps left by the plague. The exceptionally high death rate among clergy was to some extent overcome by throwing open to laymen posts usually reserved for monks or priests. Many estates, left without heirs, were taken over by the City Council. By 1353, a balanced budget had almost been achieved. What was left of the old oligarchy gained enormously through inheritances from their dead relations and the accumulation of power in fewer hands. It seemed that the *status quo ante* had been restored, indeed that the old order was even more firmly established than before the plague.

But the gloss of normality was quickly cracked. The remnants of the oligarchy had not been the only group to profit financially from the epidemic. A class of new rich arose and wished to play the part in the city's government to which they felt the length of their purse entitled them. But their pretensions met with a chilly response. No concessions were made to meet them and harsh sumptuary laws were passed to curb the ambitions of those who affected the trappings of higher station than their birth and education justified. Meanwhile the poor, among whom the disease had raged the worst, often found that they had lost even the little which they had once possessed. The gap which divided them from their luckier neighbours grew ever wider.

By the time of the Black Death the Government of Nine had ruled Siena without serious challenge for some seventy years. A few years later it seemed successfully to have weathered the storm and to have launched Siena on another era of stable prosperity. Yet in 1354 it fell. It can be argued that this was not a direct consequence of the plague but, equally, it is certain that the Black Death, in Dr Bowsky's phrase, 'was instrumental in creating demographic, social and economic conditions that greatly increased opposition to the ruling oligarchy.'[32] Without some such prior conditioning it is hard to see how the necessary force and will to overthrow the oligarchy could have sprung into life. It is not desirable, at this point in the narrative, to give much attention to the long term effects of the Black Death on the society which it had devastated. But it is important to bear in mind the lesson of Siena: that a patient has not necessarily recovered because his more obvious wounds are healed.

By the winter of 1348, about a year after its first appearance in Sicily, the Black Death in Italy was past the worst. There were to be minor outbreaks in the next year or two and it was to be much longer before the man in the street felt himself entirely safe; he barely did so, in fact, before the next epidemic was upon him in the early 1360s. But the period of acute crisis was over. Pope Clement VI threatened to revive the danger when he yielded to pressure from many countries and proclaimed 1350 a Holy Year. The first Jubilee had been held in 1300 and it had not been intended to hold another until a century later but, in the circumstances, the Pope agreed to advance the date and to grant special indulgences to all who made the journey to Rome. To fill the roads of Europe with wandering pilgrims and concentrate them in the heart of one of the areas worst struck by plague could well have been the surest means of renewing the full force of the epidemic. Matteo Villani, one of the sounder of the chroniclers when it came to statistics, wrote that around Easter, though the pilgrims were too numerous to count, there must have been more than a million visitors to Rome.[33] The figure must be by far too large but the influx of pilgrims from all over Europe was certainly immense.

St Bridget of Sweden was among the visitors, arriving early in 1349 when the Black Death was still a lively menace. She had clear views about the proper method of tackling the epidemic: 'abolish earthly vanity in the shape of extravagant clothes, give free alms to the needy and order all parish priests to celebrate Mass once a month in honour of the Holy Trinity.'[34] These rather humdrum measures do not appear greatly to have impressed the Romans but she still scored a considerable personal success. One male Orsini, it is recorded, had caught the plague and was despaired of by the doctors. 'If only the Lady Bridget were here!' sighed his mother. 'Her touch would cure my son.' At that moment in walked the saint. She prayed by the invalid's bedside, laid her hand on his forehead and left him, a few hours later, fully restored to health.

St Bridget's attentions were not much needed. In spite of the Pope's ill-judged decision, Holy Year brought little in the way of fresh outbreaks. But the damage was already bad enough. Italy had been depopulated. But when one tries to describe this dramatic concept in slightly more mathematical terms, the difficulties begin. On the basis of our present knowledge it is quite impossible to put forward even the most approximate figure and state with authority that such a proportion of Italy's peoples must have died. Even in England, with its wealth of ecclesiastical and civil records and its army of diligent scholars, only a more-or-less informed guess is feasible. A fortiori in Italy, where many regions have been the subject of little or no research if indeed the materials for such research exists, an overall estimate has little value. Sometimes it is possible to fix a movement of population over a longer period. It is, for instance, reasonably well established that the population around Pistoia in 1404 was only some 30% of what it had been in 1244.[35] But the data does not exist which would enable one to pinpoint the proportion of this decline to be attributed to the year 1348.

But the fact that any estimate for the whole of Italy must be highly speculative does not preclude a guess. Doren, in his *Economic History*,[36] has estimated that between 40% and 60% of town dwellers died and that, in the countryside, the proportion must have been much lower. Figures like these cover a multitude of qualifications. In Tuscany, for example, where the plague was exceptionally severe, more peasants died than in certain cities which escaped lightly, such as Milan or Parma. For some areas, where no statistics whatsoever can be garnered, the only remedy is to apply the proportion established for roughly similar parts of the country and hope for the best. A shot in the dark, or at least the twilight, however tentative, is still better than nothing. If one assumes that a third or slightly more of Italy's total population perished, it is unlikely that one would be very badly wrong and certain that nobody could prove one so.

FOUR

France: the State of Medical Knowledge

The Black Death seems to have arrived in France only a month or two after its first outbreak on the mainland of Italy; according to an anonymous Flemish cleric in one of those same ill-fated galleys which had been expelled from Italy towards the end of January 1348.[1] The galley called first at Marseilles, from where it was chased, rapidly but still not rapidly enough, by the horrified authorities. Thence it continued its destructive course, spreading the plague to Spain and leaving a trail of infection along the coast of Languedoc.

France was certainly one of the most populous and should also have been the most prosperous country of Europe. Professor Lot has put its population in 1328—the population, that is to say, of the France of its present frontiers, including part of Flanders, Burgundy, Brittany and Guienne—at between twenty-three and twenty-four million.[2] Professor Renouard estimates the total twenty years later at somewhere near twenty million.[3] The density of population in the countryside was more or less what it is to-day:[4] a burden which the land was hard put to it to support since production per acre was barely a third of the present figure. On the whole the latest studies tend to indicate a total population somewhat lower[5] than earlier estimates but no one would deny that the rural population was dense by medieval standards, comparable with that of Tuscany, and that the pressure of population on resources was fast becoming intolerable.

Left to itself the French countryside was probably more capable of supporting such a crowd than any other region in Europe. But where Italy had to endure its Guelphs and Ghibellines, France had the English. King Edward III had no intention of leaving France to itself or, to look at the matter in a more chauvinistic light, was justifiably outraged by French interference

with his Duchy of Guienne and their support for David Bruce in
Scotland. John of Bridlington, indeed, ascribed the plague in
France mainly to the contumacious policy of its King.[6] He
pointed out that the French had been guilty of avarice, luxury,
envy, gluttony, anger, sloth and conspicuous lack of devotion to
the saints but that the chief crime which had called down divine
vengeance was undoubtedly the failure of Philip VI to allow
Edward III free and peaceful enjoyment of his inheritance. He
did not go on to explain the curious circumstance that God had
subsequently extended the scope of his wrath to embrace the
virtuous and ill-used English.

Whether or not the policy of Philip VI provoked the plague,
it certainly led to the Hundred Years War between France and
England; in the long run to the detriment of both countries, in
the short with disastrous consequences for his own. From 1337,
when Philip VI announced that the English throne had forfeited
Guienne and Edward III retorted by claiming the throne of
France, the French peasant, in great areas of his country, no longer
knew the meaning of the word security. A brief truce after the
naval battle of Sluys quickly ended in renewed warfare. In 1346
Edward III landed in Normandy with some 15,000 men. On
25 August he won a crushing victory at Crécy. The subsequent
siege of Calais lasted a year. Military casualties in the campaign,
by modern standards or when viewed against the size of the
French population, were insignificant but the damage to civilian
morale and to the agricultural richness of the country was im-
measurable. To the luckless villagers, whose few possessions had
alternately been looted by French or English soldiery, the apparition of the plague seemed merely the culminating phase in a
process designed by God to end in their total destruction.

Within a month, wrote one authority, fifty-six thousand people in
Marseilles met their end.[7] The figure seems improbably high but,
as in many sea-ports where bubonic and pulmonary plague raged
side by side, mortality was greater than in the inland regions.
From the Mediterranean the epidemic advanced along two main
lines. To the west it quickly reached Montpellier and Narbonne.
It afflicted Carcassonne between February and May; moved on to
Toulouse and Montauban and finally reached Bordeaux in

August. To the north, Avignon was attacked in March, April and May; Lyons in the early summer, Paris in June and Burgundy in July and August. Flanders was exempt until 1349.

In Perpignan the plague took much the same course as in Avignon though, as was usually the case, it passed more quickly in the smaller city. The disruption of everyday commercial life is shown strikingly by statistics of loans made by the Jews of Perpignan to their Christian co-citizens. In January, 1348, there were sixteen such loans, in February, twenty-five, March, thirty-two, eight in the first eleven days of April, three in the rest of the month and then no more till 12 August. Of 125 scribes and legists known to have been active shortly before the Black Death only forty-five appear to have survived—even with a reduction for natural mortality a death rate of between fifty and sixty percent seems likely. Physicians fared even worse—only one out of eight surviving—while sixteen out of eighteen barbers and surgeons perished or, at least,[8] disappeared.

'Laura,' wrote Petrarch in his manuscript of *Virgil*, 'illustrious by her virtues and long celebrated in my songs, first greeted my eyes in the days of my youth, the 6th of April, 1327, at Avignon; and, in the same city, at the same hour of the same 6th of April, but in the year 1348, withdrew from life, whilst I was at Verona, unconscious of my loss . . .

'Her chaste and lovely body was interred on the evening of the same day in the Church of the Minorites: her soul, as I believe, returned to heaven whence it came.

'To write these lines in bitter memory of this event and in the place where they will most often meet my eyes has in it something of a cruel sweetness, but I forget that nothing more ought in this life to please me, which, by the grace of God, need not be difficult to one who thinks strenuously and manfully of the idle cares, the empty hopes and the unexpected end of the years that are gone . . .'[9]

Avignon in 1348 had been for nearly half a century the seat of the Popes. As such it had swollen from an always considerable town to one of the great cities of Europe. Its role as papal capital ensured that it would be one of the most visited centres of Christendom; an easy prey for a plague that thrived on every kind of social intercourse. An unnamed canon writing to a friend in Bruges spoke of half the population of Avignon being dead, seven thousand houses shut up and deserted, eleven thousand corpses buried in six weeks in a single graveyard, sixty-two thousand victims in the first three months of the epidemic.[10] Another record put the total of the dead at more than a hundred and twenty thousand[11] while the German historian, Sticker, on still less certain authority, even ventured as far as a hundred and fifty thousand.[12] It is, at least, not hard to believe that half the population died though one of the few verified facts might be taken as indicating a lower figure. The Rolls of the Apostolic Chamber show that only ninety-four out of four hundred and fifty, or 21%, of the members of the Papal Curia died during the Black Death.[13] But this is not much of a pointer to the overall death rate. Nobody would have expected the well-fed and well-housed senior staff of the papal establishment to perish at the rate of their fellow mortals.

On the whole the churchmen of Avignon seem to have behaved creditably during the plague; churchmen in the widest sense that is, from papal councillor to penniless and itinerant monk. 'Of the Carmelite friars at Avignon,' wrote Knighton uncharitably,[14] 'sixty-six died before the citizens knew the cause of the calamity; they thought that these friars had killed each other. Of the English Austin Friars at Avignon not one remained, nor did men care.' Knighton had all the contempt of a Canon Regular for these turbulent and often embarrassing colleagues. 'At Marseilles, of one hundred and fifty Franciscans, not one survived to tell the tale; and a good job too!' was another of his still harsher comments. Yet in fact there is no reason to doubt that the mendicant orders behaved at Avignon with as much courage and devotion as they did elsewhere and that their reputation rose accordingly.[15]

Pope Clement VI himself played a slightly less forthright part. There is no doubt that he was preoccupied by the horrors of the plague and genuinely disturbed and distressed for his people. Though by no means celebrated as an ascetic he was good-hearted and honourable, anxious to do what was best for his flock. He did all he could to ease the path of the afflicted by relaxing the formalities needed to obtain absolution and ordered 'devout processions, singing the Litanies, to be made on certain days each week'. Unfortunately such processions tended to get out of hand; at some, two thousand people attended, 'amongst them, many of both sexes were barefooted, some were in sack cloth, some covered with ashes, wailing as they walked, tearing their hair, and lashing themselves with scourges even to the point where blood was drawn.'[16] At first the Pope made a habit of being present at these processions, at any rate when they were within the precincts of his palace, but excesses of this kind revolted his urbane and sophisticated mind. He also realised that large concourses, attended by the devout from all over the region, were a sure means of spreading the plague still further, as well as providing a breeding ground for every kind of hysterical mob outburst. The processions were abruptly ended and the Pope from then onwards sought to discourage any kind of public demonstration.

Not unreasonably, Pope Clement VI calculated that nothing would be gained by his death and that, indeed, it was his duty to his people to cherish them as long as possible. He therefore made it his business to stay alive. On the advice of the papal physician, Gui de Chauliac, he retreated to his chamber, saw nobody, and spent all day and night sheltering between two enormous fires. For a time he took refuge in his castle on the Rhône near Valence but by the autumn he was again at his post in Avignon. It does not seem that the Black Death died out in the papal capital much before the end of 1348.

'Fish, even sea fish, are commonly not eaten' the horrified clerics of Bruges heard from their compatriots at Avignon, 'as people say that they have been infected by the bad air. Moreover, people do not eat, nor even touch spices, which have not been kept a year, since they fear that they may have lately arrived in the aforesaid ships. And, indeed, it has many times been observed that those who have eaten these new spices and even some kinds of sea fish have suddenly been taken ill.'

As the Black Death moved across Europe it was inevitable that a host of theories would be generated on the best methods of avoiding, preventing and curing the disease. The growing threat to France induced King Philip VI to appeal to the Medical Faculty at Paris to prepare a considered report on the subject. Their response[17] provided the most prestigious, though neither the best informed nor the most intelligent, of the many studies of

the Black Death in action. The plague literature as a whole, drawn
from some half-dozen countries, was voluminous, repetitious and
of little value to the unfortunate victims of the epidemic. Before
considering it, however, it is worth taking a quick look at the
growth of medical knowledge before and during the Middle
Ages so that the disadvantages and limitations under which the
medieval physician laboured can be better understood.

Modern medical science, if a gross over-generalisation may be
forgiven, began with Hippocrates. It was he who first conceived
ill health, not as a series of unrelated and essentially inexplicable
catastrophes but as an orderly process calling on each occasion
for examination of symptoms, diagnosis of malady and prescription of cure. For any study of the Black Death his importance
is paramount since he was the first student of epidemiology and
the first to distinguish between epidemic and endemic diseases. In
his First and Third Books of Epidemics and the four volumes of
notes compiled either by Hippocrates himself or by his son he
set out to analyse the factors which led to a disease settling in any
given area and becoming endemic. The next stage was to define
and explain the causes, climatic, meteorological or latent within
the body of man himself, which provoked a subsequent epidemic
outburst. It was his particular concern to work out a relationship
between each type of epidemic and the different environmental
conditions in which it flourished. This 'katastasis,' as he called it,
was, it seemed to him, best established on astronomical evidence
—a red herring which Hippocrates himself might in time have
transcended but which was to bedevil medical research for many
centuries.

The main flaw in the monumental labour of Hippocrates was
that he had insufficient data from which to draw valid conclusions.
He deduced, for instance, that spells of warm, moist weather were,
in themselves, conducive to ill health; a thesis reasonable enough
where malarial regions were concerned but irrelevant if not
positively misleading when applied to England. His great
achievement was to have provided a blue-print for research on
which subsequent generations should have worked. The tragedy
is that the vast compilation of case histories, on which a serious
study of epidemiology could alone have been based, was not made
by his successors. After the death of Hippocrates in 377BC, medical science slumbered for five hundred years; it awoke only to

find itself rigidified by the misplaced formalising genius of Galen of Pergamos.

Galen was one of the outstanding intellects of his age and a great experimental physiologist. But, when it came to epidemiology, rather than work from the Hippocratic base and accumulate fresh data from which empirically to establish new and constructive theses, he instead elected to devise an inflexible theoretical pattern which left no room for further research or original thought. He lived through a major epidemic of bubonic plague but the phenomenon was in no way reflected in his work. To attempt to summarise Galen's complicated and, within his own terms of reference, logically faultless theorising would be to reduce it to a parody. Suffice it to say that he believed ill health to depend on the interaction of temperament, the constitution of the atmosphere and certain other factors such as excessive or ill-judged eating and drinking. Temperament and constitution in their turn depended on the blending of the elementary qualities and any failure to achieve perfect balance led to one of a number of possible discords. The permutations on these factors were developed into an intricate mathematical pattern: a computer into which the details of any case could be fed and a logically satisfactory explanation provided.

Unfortunately, though the logic might be impeccable, its relevance to anything so mundane as the prevention or cure of plague was sadly tenuous. What was worse, the medieval physician believed that Galen had said the last word on epidemics and that any further research was unnecessary if not positively disrespectful to the teachings of the master. And yet the teachings of the master themselves were in doubt since the original texts had been largely lost and doctors in the West for several centuries worked almost exclusively from inadequate Latin versions of Arabic translations of Hippocrates and Galen. The result was an Arabic-Latin literature, in Dr Singer's description:

'. . . generally characterised by the qualities most often associated with the words "medieval" and "scholastic." It is extremely verbose and almost wholly devoid of the literary graces. An immense amount of attention is paid to the mere arrangement of the material, which often occupies its author more than the ideas that are to be conveyed. Great stress is laid on argument, especially in the form of syllogism, while observation of nature is entirely in the background . . . Lip-service is often paid to Hippocrates, but his spirit is absent from these windy discussions.'[18]

Nurtured on such material it is hardly surprising that medical science did not flourish in the Middle Ages. 'The Dark Ages for Medicine,' wrote Dr Singer, 'began at the death of Bede in 753.'[19] They did not end until long after the Black Death had run its course. But the failings of the fourteenth-century doctors should not be exaggerated nor their limitations presented as grotesque extravagances. Ill-informed and unimaginative they might have been but there was, on the whole, surprisingly little of the:

> 'Watres rubifiying, and boles galle,
> Arsenyk, sal armonyak, and brymstoon,
> And herbes koude I telle eek many oon,
> As egremoyne, valerian, and lunarie.'*

which were the stock-in-trade of Chaucer's alchemist.

The situation of medicine was not helped by the stern determination of the medieval churchman to keep the physician in his place. What Professor Gurlt described as 'that fatal exaggeration which enthroned theology not merely as mother but as Queen of all the sciences,'[20] ensured that the doctor would play a secondary role. In the sick room it was the priest who took the lead and the doctor who humbly offered his services once the praying was over. Before he even treated a patient the doctor was supposed to establish whether he had first confessed; if he had not, then medicine would have to wait its turn. Sometimes the doctor would manage to assert himself but, in general, the more eminent the invalid, the more likely it was that he would find himself thrust into the back row behind a bevy of churchmen and courtiers. When the disease worked quickly a doctor might not even be admitted to his patient's bedside until death was imminent or had actually occurred.

But the Church, by the stranglehold which it had on every field of education, ensured that the invalid would have gained little

---

* For this and subsequent extracts from *The Canterbury Tales* I append as a footnote Professor Coghill's admirable rendering published by Penguin Books.

> 'Water in rubefaction; bullock's gall,
> Arsenic, brimstone, sal ammoniac,
> And herbs that I could mention by the sack,
> Moonwort, valerian, agrimony and such.'

even if the doctor had been given a freer hand. All medical teaching at the universities was on lines laid down by the Church and consisted mainly of the reading of outmoded texts with a brief and usually misleading 'interpretation' by the professor. Surgery was the poor relation of an anyhow impoverished science. In 1300, Boniface VIII published a Bull forbidding the mutilation of corpses. His object was to check the excesses of relic hunters but, incidentally, he dealt a crippling blow to would-be anatomists. Soon afterwards the Medical Faculty of Paris formally declared itself an opponent of surgery. At Montpellier, supposed to be among the most enlightened of the medical schools, there was one practical anatomy lesson every two years. This long and eagerly awaited occasion consisted merely of the opening of an abdomen and a cursory exposition of its contents. It was not till the end of the fifteenth century that Sixtus IV authorised the practice of dissection and even then specific authority had to be obtained on each occasion.

Given such handicaps it would have been miraculous if the medical profession had met the Black Death with anything much more useful than awe-struck despair. Their efforts were as futile as their approach was fatalistic. Not only were they well aware that they could do little or nothing to help but they considered it self-evident that an uncharitable Deity had never intended that they should. 'The plague,' wrote Gui de Chauliac, one of their most distinguished and, incidentally, successful practitioners, was 'shameful for the physicians, who could give no help at all, especially as, out of fear of infection, they hesitated to visit the sick. Even if they did they achieved nothing, and earned no fees, for all those who caught the plague died, except for a few towards the end of the epidemic who escaped after the buboes had ripened.'[21] A doctor not prepared to visit the sick must, of course, labour under a singular disadvantage but de Chauliac was certainly right in his contention that, from the point of view of the infected, it made little difference. Nothing in the medical literature which survives suggests that the treatment of the doctors, though it may sometimes have eased a patient's sufferings, can have been directly responsible for a single cure.

The views and activities of the doctors are reasonably well-known through the plague tractates which they left behind them. Sudhoff's *Archives*[22] already reproduce well over two hundred and eighty of these. Many relate to other phases of the great pandemic but seventy-seven were written before 1400 and at least twenty before 1353. The majority of the most important studies relating to the Black Death have been analysed by Dr Anna Campbell in her invaluable work, *The Black Death and Men of Learning*.[23] The preamble to the Report of the Faculty of Medicine at Paris demonstrates admirably the vague, hopeless search for arcane solutions which appeared again and again in the tractates:

'Upon seeing effects whose cause is concealed even from the most highly trained of intellects, the mortal mind must ask itself, especially as there is in it an innate desire for appreciation of the good and of the true, for what reason everything seeks good and desires knowledge . . .'

But the windy nothingness of this somewhat unhelpful speculation was by no means typical. There was much shrewd observation and a certain amount of common sense and sound judgement in the recommendation of measures which, though no more than palliatives, still did more good than harm. Certainly there was a depressing readiness to stress that flight was the only possible defence against the plague or to argue that, if flight were impossible, there had better be immediate recourse to prayer. But the patient was also given a certain amount of guidance on how he should conduct himself. It seems unlikely that much confidence was inspired among those threatened or afflicted by the plague but some people must at least have been given a ray of hope and a feeling that they were not entirely helpless in the face of destiny.

Differences of opinion between the experts were frequent. Simon of Covino[24] considered that the pregnant woman and, even more, 'those of fragile nature' would succumb and that no doctor could help them. The undernourished pauper would be the first to go—a reasonable conclusion which the Medical Faculty, however, rejected, claiming that those 'whose bodies are replete with humours' were the most vulnerable. Ibn Khātimah[25] agreed with the Faculty. People of 'hot, moist temperament' were the most exposed; the ultimate in peril was to be a stout young woman with a taste for lechery.

There was more agreement on the best place to live so as to avoid the plague. Seclusion, obviously, must be a first priority. After that, the problem was how to keep out of the way of the dreadful miasma, the infected air which, borne generally by the south wind, carried death from land to land. A low site, sheltered from the wind, was of course desirable. The coast was to be shunned; with good reason because of the threat from ship-borne rats though the danger was visualised by the tractators as corrupt mists creeping lethally across the surface of the sea. Marsh lands were not to be recommended for here too rose the killing mist. Houses should face north; windows preferably be glazed or covered with waxcloth.

Even in the most prosperous practices—and where there was no prosperity there was generally no doctor—the doctor recog-nised that many of his clients would not be able to flee to remote spots where they might hope to escape the plague. Rules had to be laid down for the conduct of life in a plague-stricken area. If infection was carried by corrupted atmosphere then what was needed was something to build up counter-bodies within the air. Dry and richly scented woods were to be burnt: juniper, ash, vine or rosemary. Anything aromatic was of value: wood of aloes, amber, musk, or, for the less prosperous, cyprus, laurel and mastic. A typical recipe for a powder to throw on the fire[26] was one ounce each of choice storax, calamite and wood of aloes mixed in a mortar with rose-water of Damascus and made into small, oblong briquettes. The house was to be filled, whenever possible, with pleasant smelling plants and flowers and the floors sprinkled with vinegar and rose-water. If one was unfortunate enough to have to leave one's house a prudent precaution would be to carry an amber or smelling apple. If amber was too expensive a cheaper but efficacious substitute could be contrived from equal portions of black pepper and red and white sandal, two portions of roses and a half portion of camphor. The resulting blend was then pounded for a week in a solution of rose-water and moulded into apples with paste of gum arabic.

It is doubtful if these precautions did anything to reduce the risk of infection with the possible exception of one recipe of Dionysius Colle for a powder, used for throwing on the fire, which contained sulphur, arsenic and antimony.[27] The first of these is now recognised as being destructive to bacteria as well as to rats and fleas. The other compounds, in most cases, had the minor merit of making the usually smelly medieval house a little more agreeable to live in.

Vegetable inactivity was the ideal posture in which to meet

the plague. If one had to move, at least move slowly; exercise introduced more air into the body and, with the air, more poison. Hot baths, which opened pores in the skin, were to be shunned for the same reason, though it was beneficial to bathe the face and hands from time to time in vinegar or the inevitable rose-water.

There was general agreement also on the best kinds of preventive medicine. A fig or two with some rue and filberts taken before breakfast was a useful start to the day. Pills of aloes, myrrh and saffron were popular. One authority[28] placed his confidence in ten-year-old treacle blended with some sixty elements, including chopped-up snakes, and mixed with good wine. Rhubarb and spikenard was a compound easier to manufacture and to swallow. Witchcraft joined herbalism in the works of Gentile of Foligno[29] who recommended powdered emerald; a remedy so potent that, if a toad looked at it, its eyes would crack. Gentile also suggested etching on an amethyst the figure of a man bowing, girded with a serpent whose head he held in his right hand and whose tail in his left. To be fully operative the stone had first to be set in a gold ring.

Bleeding was generally held to be a useful preventive device; Ibn Khātimah,[30] for instance, feeling that it could only be beneficial to lose up to eight pounds. Diet was important. Anything which quickly went bad in hot weather was to be avoided. So was fish from the infected waters of the sea. Meat should be roast rather than boiled. Eggs were authorised if eaten with vinegar[31] but should never be taken hard-boiled.[32] Anyone trying to follow the advice of every expert would have been sadly perplexed. Ibn Khātimah approved of fresh fruit and vegetables but no one else agreed. Gentile of Foligno recommended lettuce, the Faculty of Medicine at Paris forbade it. Ibn Khātimah had faith in egg plant, another expert deplored its use.

It was bad to sleep by day or directly after meals. Gentile believed that it was best to keep steady the heat of the liver by sleeping first on the right side and then on the left. To sleep on one's back was disastrous since this would cause a stream of superfluities to descend on the palate and nostrils. From thence these would flow back to the brain and submerge the memory.

Bad drove out bad and a school of thought maintained that to imbibe foul odours was a useful if not infallible protection. According to John Colle: 'Attendants who take care of latrines and

those who serve in hospitals and other malodorous places are
nearly all to be considered immune.' It was not unknown for
apprehensive citizens of a plague-struck city to spend hours each
day crouched over a latrine absorbing with relish the foetid
smells.

A tranquil mind was one of the surer armours against infection.
Ideally one should retreat to Boccaccio's enchanted glade, live
beautifully, pass one's time in dalliance and in practising the art of
conversation. But dalliance should not be carried too far: sex,
like wrath, heated the members and disturbed the equilibrium.
One's mind should be resolutely closed to the agonies of one's
fellow men; sadness cooled the body, dulled the intelligence and
deadened the spirit.

It seems unlikely that the intelligent and enlightened men who
worked out these preventive measures had any great faith in their
efficacy. Essentially they were a morale-building exercise: the
morale of the physician, in that they made him feel at least remotely in control of the situation, and of the patient, in that they
offered a slight hope of escape from death. But if the doctors
lacked confidence in their capacity to keep the plague at bay, still
more did they doubt their ability to cure it once it had struck.
They knew too well how few of the sick recovered. But this
knowledge of their helplessness did not stop them putting forward a host of remedies.

Bleeding was an even more important part of the cure than it
had been of prevention. The blood that emerged from the infected would normally be thick and black; it boded even worse
for the victim if a thin green scum rose to the surface. If the
patient fainted, instructed Ibn Khātimah somewhat heartlessly,
pour cold water over him and continue as before. Most surgeons
bled for the sake of bleeding, not worrying much where the
incision was made. John of Burgundy[33] was more scientific. He
believed in the existence of emunctories, from which the poison
could be expelled by bleeding. The evil vapours, having entered
by the pores of the skin, were carried by the blood either to the
heart, the liver or the brain. 'Thus, when the heart is attacked, we
may be sure that the poison will fly to the emunctory of the heart,
which is the armpit. But if it finds no outlet there it is driven to
seek the liver, which again sends it to its own emunctory in the
groin. If thwarted there, the poison will next seek the brain,

when it will be driven either under the ears or to the throat.'
Each emunctory had a surface vein which corresponded to it and
a skilled surgeon could there intercept the poison on its devil's
progress around the body and draw it off before it did more
mischief. A common and disastrous mistake was to make the
incision on the wrong side of the body; this not only wasted good
blood but meant that healthy limbs were corrupted by the de-
graded liquid which poured in to make up the loss.

As well as bleeding, it was useful to open and cauterise the
plague boils or buboes. Various curious substances were applied
to the boils to draw off the poison. Gentile used a plaster made
from gum resin, the roots of white lilies and dried human excrement, while Master Albert was in favour of an old cock cut
through the back. Ibn Khātimah believed that an operation on
the bubo was possible between the fourth and seventh day of the
disease when the poison was flowing from the heart to the boils.
But even a slight mistake in timing could lead to the escape of the
vital principle from the heart and the immediate death of the
patient.

Various soothing potions were prescribed, in particular a
blend of apple-syrup, lemon, rose-water and peppermint. This
must at least have been pleasant to drink. Even this consolation
was removed when powdered minerals were added to the mix-
ture. There was some belief in the virtues of emeralds and pearls
and the medicinal qualities of gold were taken for granted by most
authorities. Take one ounce of best gold—was Gentile's recipe—
add eleven ounces of quicksilver, dissolve by slow heat, let the
quicksilver escape, add forty-seven ounces of water of borage,
keep airtight for three days over a fire and drink until cure or,
more probably, death supervened. At least the high price of gold
ensured that not many invalids could afford to be poisoned by
such medicines.

Just how little the doctors learned from the Black Death is
shown by the Tractate of John of Burgundy or John à la Barbe,
published in 1365.[34] It is true that the author had gained much of
his experience and no doubt devised his treatments in 1348 and
1349 but he had lived through a second great epidemic in 1361
and, at the time he wrote, was distilling what should have been a
lifetime's learning. The same sterile analysis of causes appeared,
the same catalogue of futile preventive methods and still more

futile cures. Given the state of medical learning no great leap
forward was possible but if Hippocrates had been alive he would
at least have discarded a lot of dead wood of proven uselessness
and made some sensible and valuable deductions about the conditions in which the epidemic seemed to thrive and the best means
of removing them. Of this there was nothing; only the regurgita-
tion of long discredited dogmas and, from time to time, the addition of some new mineral or vegetable gimmick to give the technique of the writer a flavour of modernity.

> 'I pray to God, so save thy gentil cors
> And eek thyne urynals and thy jurdones,
> Thyn ypocras and eek thy galiones,
> And every boyste ful of thy letuarie.'*

Fourteenth-century men seem to have regarded their doctor in
rather the same way as twentieth-century men are apt to regard
their priest, with tolerance for someone who was doing his best and
the respect due to a man of learning but also with a nagging and
uncomfortable conviction that he was largely irrelevant to the
real and urgent problems of their lives. They were, of course,
ready to believe almost anything which was told them with
authority but their faith had been undermined by the patent lack
of confidence on the part of the doctors themselves. Sometimes,
under intolerable stress, scepticism would give way to something more primitive and violent, tolerance would crack and the
doctor find himself execrated as the architect of the disease which
he had proved so signally ill-qualified to check. But such mo-
ments of revulsion were rare and in general the doctor preserved
a privileged position. Chaucer's mockery and the occasional
abuse of an aggrieved patient was the worst that he had to bear.

However little use the doctors may have been, the average
Parisian at least had the comfort of knowing that he was far
better provided with medical attention than his contemporaries.
There were more doctors, predominantly Jewish, in Paris than

---

* 'God's blessing on you, Doctor, not forgetting
   Your various urinals and chamber pots,
   Bottles, medicaments and cordial tots
   And boxes brimming all with panaceas.'

in any other city of Europe and all surgeons had passed an examination and had been licensed by a panel of master surgeons
trained at the Châtelet.[35] The fashionable course for study was
based on the works of the Arab surgeon Razi and an ointment
called 'Blanc de Razès' was on sale at apothecaries as a cure
recommended for virtually any ailment. But no amount of oint-
ments nor the wisdom of the august Faculty of Medicine itself
could do much to help the Parisian when the Black Death broke
about him.

It seems that the first authenticated cases of plague in the
capital were noted in May or June, 1348, though the full force of
the epidemic was not felt until several months later. It did not die
out until the winter of 1349. The chronicler of St Denis put the
death roll at about fifty thousand,[36] a surprisingly conservative
estimate for a city that may well have had more than two hundred
thousand inhabitants.[37] Certainly there is no reason to think the
figure exaggerated. An analysis based on church warden accounts
for the parish of St Germain l'Auxerrois showed that seventy-
eight legacies were bequeathed to the church between Easter,
1340 and 11 June, 1348.[38] In the next nine months the number
rose to four hundred and nineteen—a rate forty times as high. 'It
seems,' commented Mollat, 'that the plague sent men to their
lawyers as fast as to their confessors.' On this, admittedly somewhat slender basis, he calculated that the plague was at its worst
almost at the end, in September and October, 1349, and that it ran
an unusually protracted course. The Bishop, Foulque de Chanac,
died in July, 1349; the Duchess of Normandy, Bonne de Luxem-
bourg, in October and her mother-in-law, Jeanne de Bourgogne,
when the danger must have seemed almost over, on 12 December.

'There was,' wrote the best-known chronicler of the Black
Death in Paris, 'so great a mortality among people of both sexes,
of the young rather than of the old, that it was hardly possible to
bury them.'[39] This is not the only reference to the plague striking
the young rather than the old, the strong before the weak.
Statistically, as will be seen more clearly in the case of England,[40]
there seems no reason to credit such a theory. The position is, of
course, complicated by the tendency of the old to die from other
causes without a major epidemic to speed them on their way, but,
even after allowing for this, the young and fit, as was to be expected, proved the more likely to resist the disease. The death of

a strong young man is naturally more shocking and more likely to be remembered: to this, perhaps, is due the conviction on the part of certain chroniclers that his chances of death were unfairly high.

'In the Hôtel-Dieu at Paris,' continued the chronicler, 'so great was the mortality that for a long time more than five hundred corpses were carted daily to the churchyard of St Innocent to be buried.' Cardinal Gasquet has suggested that this is a misprint for fifty.[41] Certainly the latter figure is probably nearer reality but there seems little reason to doubt that William of Nangis would have preferred the larger, more imposing total. 'And those holy sisters,' he continued, 'having no fear of death, tended the sick with all sweetness and humility, putting all honour [sic ? fear] behind their back. The greater number of these sisters, many times renewed by death, now rest in peace with Christ, as we must piously believe.' Such examples of self-abnegation were rare enough to deserve special mention; for the most part, in Paris as in every other city, the rule of the day was *sauve-qui-peut*. Charity, if it was to be found at all, began at home and ended there. Even the priests 'retired through fear,' leaving the sick to shrive themselves or, if they were lucky, to gain the attention of a mendicant friar.

When the pastors set such an example it was hardly surprising that their flocks should follow suit. In the benighted city, where it seemed death to wander abroad, only those servants of death, the grave-diggers, felt themselves free to travel where they would. The rich and privileged fled, the poor remained to drown their fear in looted liquor and die in their hovels. Under the surface scum of terror and disorder ordinary decent men continued to behave in an ordinary, decent way; the life of the city fitfully continued. But to the casual visitor it must have seemed that society had disintegrated, that the plague must rage until there was not a home inviolated, not a Parisian alive. If it was not to be the end of the world, it seemed at least the end of the established order.

From Paris the plague moved northwards to the coast, which it reached in or a little before August, 1348. In this area, exceptionally, the winter checked the violence of the epidemic but with the spring it returned, evidently in its more virulent pulmonary form.

The King had fled from Paris to Normandy but the plague was
quick to follow him. At Rouen, where an over-excited contemporary calculated the dead at a hundred thousand, the Duke of
Normandy donated land for a new graveyard. At Bayeux the
Bishop and many canons died. At La Graverie, about four miles
from Vire, 'the bodies of the dead decayed in putrefaction on the
pallets where they had breathed their last.' A black flag flew above
the church as it did in all the worst affected villages of Normandy.[42]

La Léverie was a village within the parish of La Graverie. The
lady of the manor died and her relatives wished to bury her in the
churchyard. But no priest of La Graverie remained alive to conduct the service and there was no sign of a new incumbent being
appointed. The relatives appealed to the priest of neighbouring
Coulonces who was happy to bury the deceased but drew the line
at visiting La Graverie, because of the danger of infection, and
was equally reluctant to accept the corpse in Coulonces for fear
of contaminating his so far inviolate village. So this lady was
buried in the park of her own manor and the grateful relatives
arranged the transfer of La Léverie from the parish of La Graverie
to the parish of Coulonces.

St Marie Laumont in the same area lost four hundred people,
or over half the population. The epidemic there raged for three
months and had ended by September, 1348. Amiens, it seems,
must have suffered a second attack or perhaps, as occasionally
happened in the larger cities, wasted away gradually over a year
or more rather than succumbed to a brief but shattering epidemic.
As late as June, 1349, the King authorised the Mayor to open a
new cemetery on the grounds that: 'The mortality . . . is so marvellously great that people are dying there suddenly, as quickly
as between one evening and the following morning and often
quicker than that.'[43] Meanwhile the Black Death seeped across
every corner of France. Bordeaux it had reached in August, 1348,
and there caught and killed Princess Joan, daughter of King
Edward III of England, on her way to marry the son of the King
of Castile. The news of her death reached England at much the
same time as the plague itself.

To the North East the Black Death moved slowly on towards
Flanders and the Low Countries. 'It is almost impossible,' wrote
Gilles Li Muisis,[44] 'to credit the mortality throughout the whole

country. Travellers, merchants, pilgrims and others who have passed through it declare that they have found cattle wandering without herdsmen in the fields, towns and waste-lands; that they have seen barns and wine-cellars standing wide open, houses empty and few people to be found anywhere . . . And in many different areas, both lands and fields are lying uncultivated.'

Professor Renouard says that, though mortality in the French towns was terrifyingly high, the country escaped comparatively lightly.[45] In many areas this was certainly true but, though Li Muisis's account may well have been embellished in the interests of the picturesque, there are too many descriptions like it to accept Renouard's statement as invariably valid. At Givry, a large Burgundian village near Chalon-sur-Saône of between twelve and fifteen hundred inhabitants, the average death rate in the years before the plague was thirty each year. Between 5 August and 19 November, 1348, six hundred and fifteen people died. In Saint-Pierre-du-Soucy, a rural parish in Savoy, the number of households dwindled from one hundred and eight in 1347 to sixty-eight at the end of 1348 and fifty-five in 1349. In the seven neighbouring parishes, all farming communities with small and scattered populations, the number of households fell from three hundred and three to one hundred and forty-two.[46] In areas such as these, at least, something close to half the population must have perished.

In the summer of 1349, the Black Death reached Li Muisis's own city of Tournai. The Bishop, John de Pratis, was one of the first to die; then came a lull during which the citizens told themselves that they had been let off lightly. But by August the plague was raging with renewed strength: 'Every day the bodies of the dead were borne to the churches, now five, now ten, now fifteen, and in the parish of St Brice sometimes twenty or thirty. In all parish churches the curates, parish clerks and sextons, to get their fees, rang morning, evening and night the passing bells, and by this the whole population of the city, men and women alike, began to be filled with fear.'

The Town Council acted firmly to restore public confidence and check the collapse in moral standards which was likely, they feared, to bring down upon the city a still more ferocious measure of divine vengeance. Men and women who, although not married, were living together as man and wife were ordered either to marry at once or to break off their relationship. Swearing, playing dice and working on the Sabbath were prohibited. No bells were to be rung at funerals, no mourning worn and there were to be no gatherings in the houses of the dead. New graveyards were opened outside the city walls and all the dead, irrespective of their standing in the city or the grandeur of their family vaults, were in future to be buried there.

The measures seem to have been successful; if not in checking the Black Death then at least in raising the moral tone of the community. Li Muisis reported that the number of people living in sin dwindled rapidly, swearing and work on the Sabbath almost ended and the dice manufacturers became so discouraged with their sales-figures that they turned their products into 'round objects on which people told their Pater Nosters.' But gratifying though this must have been, the death roll was still terrifyingly high. 'It was strange,' said the chronicler that, 'the mortality was especially great among the rich and powerful.' Strange indeed, especially since he went on to comment: 'Deaths were more numerous about the market places and in poor, narrow streets than in broader and more spacious areas.' There seems no reason to doubt that in Tournai, as in every other city, the rich man who remained prudently secluded had a better chance of survival than the poor man who, like it or not, was forced to live cheek-by-jowl with his neighbours. But his chance was not necessarily a high one: '. . . no one was secure, whether rich, in moderate circumstances or poor, but everyone from day to day waited on the will of the Lord.'

The Middle Ages were described by Koyré, the great historian of science, as the period of *à peu près*. Everything is seen through a glass darkly; though here and there it may be possible to identify detail with fair certainty, for the most part a bold, impressionistic picture is the best that can be hoped for. No more in France than in Italy can it be said what proportion of the population died. Gui de Chauliac spoke of three quarters, other chroniclers of half; Professor Renouard, who has studied the subject as closely as anyone, can do no better than estimate that the death rate varied from an eighth to two-thirds according to the area.

But even if such details were known to us we would still be

very far from understanding what the Black Death meant to medieval man. Some hazy outline of the reactions of the French emerges from the records of the chroniclers. Men seem to have taken refuge in frenetic gaiety. It was not only in Tournai that dice and lechery were the order of the day. 'It is a curious fact,' observed Papon,[47] 'that neither the flail of war nor of plague can reform our nation. Dances, festivals, games and tournaments continued perpetually; the French danced, one might say, on the graves of their kinsmen . . .' The standards of society were relaxed; debauchery was common; thrift and continence forgotten; the sacred rules of property ignored; the ties of family and friendship denied; let us eat and drink, for to-morrow we shall die.

It is dangerously easy to allow the prejudiced records of a handful of priestly and conservative chroniclers to delude one into a vision of Europe studded with Sodoms and Gomorrahs and echoing from end to end with the rattle of dice and the laughter of tipsy courtesans. But it would be hardly less foolish to let one's rejection of such fantasies blind one to the very real degeneracy of life during the plague. The great nobles and churchmen, the richest merchants, withdrew from the city; those who were left drank, fornicated or skulked in cellars according to their inclinations. To none of them could it have seemed likely that his life would drag on for more than a few painful weeks. With no future to await and the threat of annihilation hanging over all he cared for, how could medieval man be expected to behave with responsibility? Honesty, decency and sobriety were by no means dead but they must, at times, have been uncommonly hard to find. In Paris at least there had been something not far removed from a complete collapse of public and private morality. This was not the least of the penalties which the Black Death exacted from its victims.

FIVE

Germany: the Flagellants and the Persecution of the Jews

By 1350, the plague in France had ended or, at least, so far abated as to make possible the holding of a Council in Paris to tighten up some of the laws against heresy. But in the meantime it had moved eastwards into Germany. Central Europe was thus attacked on two sides or if, as seems probable, the Black Death also advanced by land through the Balkans, on three sides more or less simultaneously. By June, 1348, it had already breached the Tyrolese Alps and was at work in Bavaria, by the end of the year it had crept up the Moselle valley and was eating into North Germany.[1]

In Styria, which it reached in November 1348, it seems to have been especially ferocious. According to the Neuburg Chronicle[2] even the wild animals were appalled at its depredations. 'Men and women, driven to despair, wandered around as if mad . . . cattle were left to stray unattended in the fields for no one had any inclination to concern themselves about the future. The wolves, which came down from the mountains to attack the sheep, acted in a way which had never been heard of before. As if alarmed by some invisible warning they turned and fled back into the wilderness.' In Frankfurt-am-Main, where Günther Von Schwarzburg died in the summer of 1349, two thousand people perished in seventy-two days.[3] In December 1349 the first case was recorded in Cologne. Six thousand died in Mainz, eleven thousand in Munster, twelve thousand in Erfurt.[4] Nearly seven thousand died in Bremen in four parishes alone.

Vienna was visited from the spring to the late autumn of 1349. Every day, wrote Sticker, five to six hundred people died; once nine hundred and sixty perished in a single day. A third part of the population was exterminated, says one record;[5] only a third survived says another.[6] The population identified the plague as the *Pest Jungfrau* who had only to raise her hand to infect a victim. She flew through the air in the form of a blue flame and, in this guise, was often seen emerging from the mouths of the dead.[7] In Lithuanian legend the same plague maiden waved a red scarf through the door or window of a house to infect its inhabitants. A gallant gentleman deliberately opened a window of his house and waited with his sword drawn till the maiden arrived. As she thrust in the scarf he chopped off her hand. He died but the rest of the village escaped unscathed and the scarf was long preserved as a relic in the local church.[8] In some areas the plague poison was believed to descend as a ball of fire. One such ball was fortunately spotted while hovering above Vienna and exorcised by a passing bishop. It fell harmlessly to the ground and a stone effigy of the Madonna was raised to commemorate this unique victory of the city's defensive system.

The details of the daily horrors are very similar to those in the cities of Italy and France and there is no need to labour them again. One point of difference is the abnormally large number of churchmen who died during the epidemic. It seems indeed that the plague fell with exceptional violence on the German clergy; because, one must suppose in the absence of other explanation, of the greater fortitude with which they performed their duties. Conrad Eubel, basing his calculations almost entirely on German sources,[9] shows that at least thirty-five percent of the higher clergy died in this period. The figure would not be exceptionally high if it related to parish priests but becomes astonishing when it applies to their normally cautious and well-protected superiors. But so far as the monks were concerned it seems that it was not only devotion to duty which led to a thinning of their ranks. Felix Fabri[10] says that in Swabia many religious houses were deserted: 'For those who survived were not in the monasteries but in the cities and, having become accustomed to worldly ways of living, went quickly from bad to worse . . .' The monks of Auwa are said to have moved in a body to Ulm where they dissipated the monastery's treasure in riotous living.

For a variety of reasons, therefore, the German Church found itself short of personnel in 1349 and 1350. One result was a sharp increase in plural benefices. In one area, between 1345 and 1347, thirty-nine benefices were held by thirteen men. In 1350 to 1352 this had become fifty-seven benefices in the hands of twelve men.

Another was the closing of many monasteries and parish churches; a third the mass ordination of young and often ill-educated and untrained clerics. As a sum of these factors, the German Church after the Black Death was numerically weaker, worse led and worse manned than a few years before: an unlucky consequence of the losses which it had suffered by carrying out its responsibilities courageously. The many benefactions which it received during the terror ensured that its spiritual and organisational weakness was matched by greater financial prosperity, a disastrous combination which helped to make the church despised and detested where formerly it had been loved, revered or, at least, accepted. By 1350, the Church in Germany had been reduced to a condition where any energetic movement of reform was certain to find many allies and weakened opposition.

One by one the cities of Germany were attacked. As always, firm statistics are few and far between and, where they do exist, are often hard to reconcile with each other. Reincke[11] has estimated that between half and two-thirds of the inhabitants of Hamburg died and seventy percent of those in Bremen; yet in Lübeck only a quarter of the householders are recorded as having perished. Most country areas were seriously affected, yet Bohemia was virtually untouched. Graus[12] has suggested that this was due to Bohemia's remoteness from the traditional trade routes yet, in the far milder epidemic of 1380, the area was ravaged by the plague. An impression is left that Germany, using the term in its widest possible sense to include Prussia, Bohemia and Austria, suffered less badly than France or Italy, but such an impression could hardly be substantiated. The Black Death in Germany, however, is of peculiar interest since that country provided the background for two of its most striking and unpleasant by-products: the pilgrimages of the Flagellants and the persecution of the Jews.

The Flagellant Movement,[13] even though it dislocated life over a great area of Europe and at one time threatened the security of governments, did not, in the long run, amount to very much. It might reasonably be argued that, in a book covering so immense a subject as the Black Death, it does not merit considered attention. In statistical terms this might be true. But the Flagellants, with their visions and their superstitions, their debauches and their discipline, their idealism and their brutality, provide a uniquely revealing insight into the mind of medieval man when confronted with overwhelming and inexplicable catastrophe. Only a minority of Europeans reacted with the violence of the Flagellants but the impulses which drove this minority on were everywhere at work. To the more sophisticated the excesses of the Flagellants may have seemed distasteful; to the more prudent, dangerous. But to no one did they seem meaningless or irrelevant —that there was method in their madness was taken for granted even by the least enthusiastic. It is this, the fact that some element of the Flagellant lurked in the mind of every medieval man, which, more than the movement's curious nature and intrinsic drama, justifies its consideration in some detail.

Flagellation as a practice seems to be almost as old as man himself. Joseph McCabe has pursued the subject with loving detail through the ages:[14] from the Indians of Brazil who whipped themselves on their genitals at the time of the new moon through the Spartans who propitiated the fertility goddess with blood until finally he arrived at the thirteenth and fourteenth century— the 'Golden Age of Pious Flagellation.' Most of these exercises were clearly if unconsciously erotic in their nature. As such, they were far removed from the pilgrimages of the Brethren of the Cross. It would be rash to assert that the Flagellants of 1348 did not satisfy, by their self-inflicted torments, some twisted craving in their natures, but 'erotic,' in its normal sense of awakening sexual appetites, is not a word which can properly be applied to their activities.

The practice of self-scourging as a means of mortifying the flesh seems to be first recorded in Europe in certain Italian monastic communities early in the eleventh century. As a group activity it was not known for another two hundred years. At this point, in the middle of the thirteenth century, a series of disasters convinced the Italians that God's anger had been called down on man as a punishment for his sins. The idea that he might be placated if a group of the godly drew together to protest their penitence and prove it by their deeds seems first to have occurred to a Perugian hermit called Raniero. The project was evidently judged successful, at any rate sufficiently so for the experiment to be repeated in 1334 and again a few years later, when the pilgrimage was led by 'a virtuous and beautiful maid.' This last enterprise ran foul of the authorities and the maid was arrested and sentenced to be burnt at the stake. Either her virtue or her beauty, however, so far melted the hearts of her captors that she was reprieved and ultimately released.

The pilgrimage of 1260 drew its authority from a Heavenly Letter brought to earth by an angel which stated that God, incensed by man's failure to observe the Sabbath day, had scourged Christendom and would have destroyed the world altogether but for the intercession of the angels and the Virgin and the altogether becoming behaviour of the Flagellants. Divine grace would be forthcoming for all those who became members of the Brotherhood: anybody else, it was clear, was in imminent danger of hellfire. A second edition of this letter was issued in time for the Black Death by an angel who was said to have delivered it in the Church of St Peter in Jerusalem some time in 1343.[15] The text was identical with the first except for an extra paragraph specifically pointing out that the plague was the direct punishment of God and that the aim of the Flagellants was to induce God to relent.

The 'Brotherhood of the Flagellants' or 'Brethren of the Cross' as the movement was called in 1348, traditionally originated in Eastern Europe, headed, according to Nohl in a pleasant conceit for which he unfortunately fails to quote authority, by various 'gigantic women from Hungary.'[16] It is to be deplored that these heroic figures quickly faded from the scene. It was in Germany that the Flagellant movement really took root. It is hard to be sure whether this was the result of circumstances or of the nature of the inhabitants. Dr Lea suggests that the German people had had their religious sensibilities stirred by the papal interdict against Louis of Bavaria and the recent earthquakes. But, if such were the causes, there would have been quite as much reason to expect the outbreak in Italy, the original home of collective scourgings, deprived as it was of its Pope and in a mood of striking melancholia.

The actual mechanism of recruitment to the Brotherhood is still obscure but the appearance of the Flagellants on the march is well attested.[17] They moved in a long crocodile, two-by-two, usually in groups of two or three hundred but occasionally even more than a thousand strong. Men and women were segregated, the women taking their place towards the rear of the procession. At the head marched the group Master and two lieutenants carrying banners of purple velvet and cloth of gold. Except for occasional hymns the marchers were silent, their heads and faces hidden in cowls, their eyes fixed on the ground. They were dressed in sombre clothes with red crosses on back, front and cap.

Word would travel ahead and, at the news that the Brethren of the Cross were on the way, the bells of the churches would be set ringing and the townsfolk pour out to welcome them. The first move was to the church where they would chant their special litany. A few parish priests used to join in and try to share the limelight with the invaders, most of them discreetly lay low until the Flagellants were on the move again. Only a handful were so high-principled or fool hardy as to deny the use of their church for the ceremony and these were usually given short shrift by the Brethren and by their own parishioners.

Sometimes the Flagellants would use the church for their own rites as well as for the litany but, provided there was a market place or other suitable site, they preferred to conduct their service in the open air. Here the real business of the day took place. A large circle was formed and the worshippers stripped to the waist, retaining only a linen cloth or skirt which stretched as far as their ankles. Their outer garments were piled up inside the circle and the sick of the village would congregate there in the hope of acquiring a little vicarious merit. On one occasion, at least, a dead child was laid within the magic circle—presumably in the hope of regeneration. The Flagellants marched around the circle; then, at a signal from the Master, threw themselves to the ground. The usual posture was that of one crucified but those with especial sins on their conscience adopted appropriate attitudes: an adulterer with his face to the ground, a perjurer on one side holding up three fingers. The Master moved among the recumbent bodies, thrashing those who had committed such crimes or who had offended in some way against the discipline of the Brotherhood.

Then came the collective flagellation. Each Brother carried a heavy scourge with three or four leather thongs, the thongs tipped with metal studs. With these they began rhythmically to beat their backs and breasts. Three of the Brethren acting as cheerleaders, led the ceremonies from the centre of the circle while the

Master walked among his flock, urging them to pray to God to have mercy on all sinners. Meanwhile the worshippers kept up the tempo and their spirits by chanting the Hymn of the Flagellants. The pace grew. The Brethren threw themselves to the ground, then rose again to continue the punishment; threw themselves to the ground a second time and rose for a final orgy of self-scourging. Each man tried to outdo his neighbour in pious suffering, literally whipping himself into a frenzy in which pain had no reality. Around them the townsfolk quaked, sobbed and groaned in sympathy, encouraging the Brethren to still greater excesses.

Such scenes were repeated twice by day and once by night with a benefit performance when one of the Brethren died. If the details of the ceremonies are literally as recorded then such extra shows must have been far from exceptional. The public wanted blood and they seem to have got it. Henry of Herford[18] records: 'Each scourge was a kind of stick from which three tails with large knots hung down. Through the knots were thrust iron spikes as sharp as needles which projected about the length of a grain of wheat or sometimes a little more. With such scourges they lashed themselves on their naked bodies so that they became swollen and blue, the blood ran down to the ground and bespattered the walls of the churches in which they scourged themselves. Occasionally they drove the spikes so deep into the flesh that they could only be pulled out by a second wrench.'

But though, gripped as they were by collective hysteria, it is easy to believe that they subjected their bodies to such an ordeal, it is impossible to accept that they could have repeated the dose two or three times a day for thirty-three days. The rules of the Brotherhood precluded bathing, washing or changes of clothing. With no antiseptics and in such grotesquely unhygienic conditions, the raw scars left by the spikes would quickly have become poisoned. The sufferings of the Brethren would have become intolerable and it seems highly unlikely that any Flagellant would have been physically capable of completing a pilgrimage. The modern reader is forced to the conclusion that, somewhere, there must have been a catch. Possibly the serious blood-letting was reserved for gala occasions, such as that witnessed by Henry of Herford. Possibly two or three victims were designated on each occasion to attract the limelight by the intensity of their sufferings.

The Flagellants were not fakes but some measure of restraint there must have been.

Certainly there was little in their chanting intrinsically likely to lead to total self-abandonment. The celebrated Ancient Hymn of the Flagellants, even in the Latin or vernacular German, was a pitiful little dirge; as remote from ecstatic excitement as a Women's Institute Choir's rendering of 'Abide With Me':

> 'Whoe'er to save his soul is fain,
> Must pay and render back again.
> His safety so shall he consult:
> Help us, good Lord, to this result . . .
> —Ply well the scourge for Jesus' sake
> And God through Christ your sins shall take . . .
> Woe! Usurer though thy wealth abound
> For every ounce thou makest, a pound
> Shall sink thee to the hell profound.
> Ye murderers and ye robbers all,
> The wrath of God on you shall fall.
> Mercy ye ne'er to others show,
> None shall ye find, but endless woe.
> Had it not been for our contrition
> All Christendom had met perdition . . .'[19]

A slightlier livelier refrain is quoted by Nohl:

> 'Come here for penance good and well,
> Thus we escape from burning hell,
> Lucifer's a wicked wight,
> His prey he sets with pitch alight.'

but even this lacks something as a stimulant.

The Flagellant Movement, at first at least, was well regulated and sternly disciplined. Any new entrants had to obtain the prior permission of their husband or wife and make full confession of all sins committed since the age of seven. They had to promise to scourge themselves thrice daily for thirty-three days and eight hours,[20] one day for each year of Christ's earthly life, and were required to show that they possessed funds sufficient to provide 4d for each day of the pilgrimage to meet the cost of food. Abso-

lute obedience was promised to the Master and all the Brethren
undertook not to shave, bathe, sleep in a bed, change their
clothes or have conversation or other intercourse with a member
of the opposite sex.

The entrance fee ensured that the poorest members of society
were barred from the Brotherhood; the strict rules, at first at any
rate conscientiously observed, kept out the sensation-mongers
who wished only to draw attention to themselves or to give unbridled scope to their passions. In these conditions, the public were
generally delighted to receive the visits of the Flagellants and, at a
small charge, to meet their simple needs. Their arrival was an
event in the drab lives of the average German peasant; an occasion for a celebration as well as for the working off of surplus
emotion. If the plague was already rife then the visit offered some
hope that God might be placated, if it had not yet come then the
penance of the Flagellants was a cheap and possibly useful insurance policy. Without at first being overtly anti-clerical the
movement gave the villager the satisfaction of seeing his parish
priest manifestly playing second fiddle if not actually humiliated.
Ecclesiastics had no pre-eminence in the movement; indeed, in
theory, they were forbidden to become Masters or to take part
in Secret Councils, and the leaders of the movement prided
themselves upon their independence from the church establishment.

So bourgeois and respectable, indeed, did the movement at first
appear that a few rich merchants and even nobles joined the pilgrimage. But soon they had reason to doubt their wisdom. As the
fervour mounted the messianic pretensions of the Flagellants became more pronounced. They began to claim that the movement
must last for thirty-three years and end only with the redemption
of Christendom and the arrival of the Millenium. Possessed by
such chiliastic convictions they saw themselves more and more,
not as mortals suffering to expiate their own sins and humanity's,
but as a holy army of Saints. Certain of the Brethren began to
claim a measure of supernatural power. It was commonly alleged
that the Flagellants could drive out devils, heal the sick and even
raise the dead. Some members announced that they had eaten and
drunk with Christ or talked with the Virgin. One claimed that he
himself had risen from the dead. Rags dipped in the blood they
shed were treated as sacred relics. All that was lacking to give the

movement the full force of a messianic crusade was a putative Messiah. Such a figure had appeared in the thirteenth century but, though there may have been one or two local claimants, no major figure emerged on this occasion to lead the Brethren of the Cross into the Millenium.

As this side of the movement's character attracted more attention, so a clash with the Church became inevitable. Already the claim of the Masters to grant absolution from sins infringed one of the Church's most sacred and, incidentally, lucrative prerogatives. A number of dissident or apostate clerics began to secure high office in the movement and these turned with especial relish on their former masters. The German Flagellants took the lead in denouncing the hierarchy of the Catholic Church, ridiculing the sacrament of the eucharist and refusing to revere the host. Cases were heard of Flagellants interrupting religious services, driving priests from their churches and looting ecclesiastical property. Other heretics—the Lollards, the Beghards and the Cellites—made common cause with them in contesting the authority of the Catholic Church.

The parallel between the Pilgrimage of the Flagellants and the preceding 'People's Crusades' became more apparent. According to John of Winterthur, the people were eagerly awaiting the resurrection of the Emperor Frederick who was expected to massacre the clergy and break down the barriers between rich and poor.[21] This delectable vision fused in the popular mind with the apocalyptic ambitions of the Brethren. The movement took on a revolutionary character and began to direct the hostility of its audiences as much against the rich layman as the cleric. What was left of the merchants and nobles now deserted the movement in disgust, leaving the extremists free to direct its passions as they wished.

The loss of its bourgeois members in itself would probably have mattered little to the Flagellant Crusade. But as they trekked from plague centre to plague centre, often bearing infection with them to those whom they were supposed to succour, it was inevitable that many of their older members should perish, including the responsible leaders who had set the standards for the rest. To make up numbers, pilgrims were recruited less remarkable for their piety or their dour asceticism than for their failure to fit into any regular pattern of life. Bandits too discovered that a con-

venient way to enter a guarded town was to tack themselves on to
the tail of a Flagellant procession. Little by little the more respectable citizens of Europe began to look with diminished favour on
their turbulent visitors.

Up to the middle of 1349, the Flagellants had things pretty
much their own way. Central and southern Germany was their
favoured hunting ground but they spread freely over Hungary,
Poland, Flanders and the Low Countries. In March they were in
Bohemia; April, Magdeburg and Lübeck; May, Würzburg and
Augsburg; June, Strasbourg and Constance; July, Flanders.
Their numbers were formidable and their needs often strained the
resources of their hosts. A single monastery in the Low Countries
had to provide for 2,500 pilgrims in a matter of six months; in
two and a half months, 5,300 Flagellants visited Tournai; when the
crusade arrived at Constance it was even claimed that there were
42,000 men in the company. If anyone opposed them their reaction was ferocious. Mendicant friars in Tournai who objected
to their pretensions were dismissed as scorpions and Antichrists
and, near Meissen, two Dominicans who tried to interrupt a
meeting were attacked with stones and one of them killed before
he could escape.

From the start, however, a few doughty spirits had declined to
be intimidated. The magistrates of Erfurt refused entry to the
Flagellants and neither from the Brethren themselves nor from
the citizens was there any attempt to defy their ruling. Archbishop
Otto of Magdeburg suppressed them from the start. In Italy they
made little impression; perhaps the example had not been forgotten of Uberto Pallavicino of Milan who, in 1260, hearing that
a Flagellant procession was on the way, erected three hundred
gibbets outside his city. The hint was taken and the pilgrims
never came. In France they were beginning to gather popular
support when Philip VI, showing unusual determination, prevented their penetrating beyond Troyes.

According to Robert of Avesbury they arrived in London in
May (or possibly September), 1349,[22] but Walsingham, who also
records the visit, puzzlingly delays it to 1350, by which time the
movement had long been on the wane.[23] '. . . there came into
England,' wrote the latter, 'certain penitents, noblemen and
foreigners, who beat their naked bodies very sharply until the
blood ran, now weeping, now singing. Yet, as was said, they did

this too unadvisedly, since they had no licence from the Apostolic See.' Robert of Avesbury puts their numbers at more than six score, 'for the most part coming from Zealand and Holland.' They are only known to have held one ceremony in London, on the open plot in front of St Paul's. They seem to have met with indifference or even hostility and were rapidly deported as unwanted guests.

But the turning point came with the declaration of war by the Church. In May, 1348, Pope Clement VI had himself patronised ceremonies involving public flagellation within the precincts of his palace at Avignon but he took fright when he saw that he could not control the movement which he had encouraged. Left to himself he would probably have turned against them sooner, but members of the Sacred College prevailed on him to hold his hand. In mid-1349, the Sorbonne was asked for its opinion and sent to Avignon a Flemish monk, Jean da Fayt, who had studied the phenomenon in his homeland. It seems that his advice was decisive. Shortly after his arrival, on 20 October, 1349, a papal Bull was published and dispatched to the Archbishops. This was followed by personal letters to the Kings of France and England. The Bull denounced the Flagellants for the contempt of Church discipline which they had shown by forming unauthorised associations, writing their own statutes, devising their own uniforms and performing many acts contrary to accepted observances. All prelates were ordered to suppress the pilgrimages and to call on the secular arm to help if it seemed necessary.

That the Pope meant business was shown when a party of a hundred Flagellants arrived in Avignon from Basle. Clement promptly interdicted public penance and prohibited their pilgrimages under threat of excommunication. Emboldened by his example, the rulers of Europe turned on the Brethren. Manfred of Sicily threatened to execute any Flagellant who appeared in his lands;[24] Bishop Preczlaw of Breslau made threats reality and had a Master burned alive. The German prelates took up the attack with especial relish. The Flagellants were denounced from the pulpit as an impious sect and harsh penalties were threatened against any who failed to return humbly to the bosom of the Church. Even those who obeyed were likely to find themselves in trouble if they had played a prominent part in the movement and hundreds were incarcerated, tortured or executed. In 1350, many Flagellants were in Rome enjoying a busman's penance by being beaten in front of the High Altar of St Peter's.

The Brethren of the Cross 'vanished as suddenly as they had come, like night phantoms or mocking ghosts.'[25] The movement did not die, indeed it was still to be encountered in the fifteenth century, but, as a threat to society or an additional headache to those grappling with the problems of the Black Death, it had effectively ceased to exist.

It is easy to poke fun at these misguided fanatics. Their superstitions were ridiculous, their practices obscene, their motivation sometimes sinister. But before condemning them one must remember the desperate fear which drove the Flagellants into their excesses. These were men who put themselves to great pain and inconvenience; in part, certainly for the sake of their own souls and their own glory, but in part also in the hope that their sacrifice might induce God to lift from his people the curse that was destroying them. There were few saints among them but, on the whole, they were not bad men. And it is impossible not to feel some sympathy for the person who, when disaster threatens, tries to do something to oppose it, however futile, instead of waiting, in abject despair, for death to strike him down.

They did achieve something. In some at least of the towns they visited they brought about a spiritual regeneration, ephemeral, no doubt, but still real while it lasted. Adulterers confessed their sins, robbers returned stolen goods. They provided some diversion at the places along their route and left behind them a fleeting hope that their pain might bring an end to the greater sufferings of the plague-stricken. But when the Flagellants had passed, often leaving new centres of infection in their wake; when the miracles did not happen, the sick did not recover, the plague did not pass; then the condition of those they left behind them must have been even worse than before they came. On the whole they probably did more harm than good.

One thing at least it is hard to forgive. In his Bull condemning them, Pope Clement VI complained that 'most of them . . . beneath an appearance of piety, set their hands to cruel and impious work, shedding the blood of Jews, whom Christian piety accepts and sustains'. The persecution of the Jews during the Black Death deserves special attention. The part which the Flagellants played in this repugnant chapter was only occasionally of the first importance but it was none the less barbarous for that.

When ignorant men are overwhelmed by forces totally beyond their control and their understanding it is inevitable that they will search for some explanation within their grasp. When they are frightened and badly hurt then they will seek someone on whom they can be revenged. Few doubted that the Black Death was God's will but, by a curious quirk of reasoning, medieval man also concluded that His instruments were to be found on earth and that, if only they could be identified, it was legitimate to destroy them. What was needed, therefore, was a suitable target for the indignation of the people, preferably a minority group, easily identifiable, already unpopular, widely scattered and lacking any powerful protector.

The Jews were not the only candidates as victims. In large areas of Spain the Arabs were suspected of playing some part in the propagation of the plague. All over Europe pilgrims were viewed with the gravest doubts; in June, 1348, a party of Portuguese pilgrims were said to be poisoning wells in Aragon and had to be given a safe conduct to get them home.[26] In Narbonne it was the English who were at one time accused.[27] But it was the leper who most nearly rivalled the Jew as popular scapegoat. The malign intentions of the leper had long been suspected by his more fortunate fellows. In 1346, Edward III decreed that lepers were no longer to enter the City of London since:

'. . . some of them, endeavouring to contaminate others with that abominable blemish (that so, to their own wretched solace, they may have the more fellows in suffering) as well in the way of mutual communication, and by the contagion of their polluted breath, as by carnal intercourse with women in stews and other secret places, detestably frequenting the same, do so taint persons who are sound, both male and female, to the great injury of the people dwelling in the city . . .'

But it is one thing to try to infect others with one's own disease for the sake of the extra companionship, another to spread the plague out of sheer devilry. When in Languedoc, in 1321, all the lepers were burnt on suspicion of poisoning wells, it was claimed that they had been bribed to do so by the Jews who, in their turn,

were in the pay of the King of Granada.[28] There were one or two
cases, notably in Spain, where lepers suffered during the Black
Death on suspicion of complicity but there do not seem to be any
where the Jews were not accorded the leading role and the lepers
cast as the mere instruments of their wickedness.

One reason for this was that nobody had cause for envying the
lepers or economic reason for wishing them out of the way. It
was very different with the Jews whose popular image was that of
the Prioress's Tale:

> '. . . sustened by a lord of that contree
> For foule usure and lucre of vileynye,
> Hateful to Christ and to his compaignye.'*

In Germany, and to some extent also in France and Spain, the
Jews provided the money-lending class in virtually every city—
not so much by their own volition as because they had been
progressively barred from all civil and military functions, from
owning land or working as artisans. Usury was the only field of
economic activity left open to them; an open field, in theory at
least, since it was forbidden to the Christian by Canon Law. In
cities such as Strasbourg they flourished exceedingly and profited
more than most during the economic expansion of the thirteenth
century.[29] But the recession of the fourteenth century reduced
their prosperity and the increasing role played by the Christian
financiers, in particular the Italian bankers, took away from them
the cream of the market. In much of Europe the Jew dwindled to
a small money-lender and pawnbroker. He acquired a large
clientele of petty debtors so that every day more people had cause
to wish him out of the way. 'It can be taken for granted,' wrote
Dr Cohn, 'that the Jewish money lenders often reacted to insecurity and persecution by deploying a ruthlessness of their
own.'[30] It is fair to criticise the medieval Jews for exacting exorbitant rates of interest from their victims but it is also only fair
to remember the extreme precariousness of their business, dependent on the uncertain protection of the local ruler and with
virtually no sanctions at their disposal to enable them to recover
their money from a reluctant debtor. To ensure their own safety

---

* '. . . . . supported by the Crown
  For the foul lucre of their usury,
  Hateful to Christ and all his company.'

the luckless Jews were forced to pay ever larger bribes to the authorities and, to raise the money for the bribes, they had to charge higher interest and press their clients still more harshly. Animosity built up and, by the middle of the fourteenth century, Shylock had been born. The Jew had become a figure so hated in European society that almost anything might have served to provoke catastrophe.

But though the economic causes for the persecution of the Jews were certainly important it would be wrong to present them as the only, or even as the principal reason for what now happened. The Jew's role as money-lender predisposed many people to believe any evil which they might hear of him but the belief itself was sincere and had far deeper roots. The image of the Jew as Antichrist was common currency in the Middle Ages. It seems to have gained force at the time of the First Crusade and the Catholic Church must accept much of the responsibility for its propagation. The vague enormity of such a concept was quickly translated into terms more comprehensible to the masses. In particular the more irresponsible priests spread rumours that the Jews kidnapped and tortured Christian children and desecrated the host. They were represented as demons attendant on Satan, portrayed in drama or in pictures as devils with the beards and horns of a goat, passing their time with pigs, frogs, worms, snakes, scorpions and the horned beasts of the field. Even the lay authorities seemed intent on fostering public belief in the malevolence of the Jews; in 1267, for instance, the Council of Vienna forbade purchases of meat from Jews on the ground that it was likely to be poisoned.

To-day such fantasies seem ludicrous. It is hard to believe that sane men can have accepted them. And yet Dr Norman Cohn[31] has drawn a revealing parallel between anti-Semitism in the fourteenth century and under the Third Reich. On 1 May, 1934, *Der Stürmer* devoted a whole issue to alleged murders of Christian children by the Jews; illustrating its text with pictures of rabbis sucking blood from an Aryan child. Most Germans were no doubt revolted by such vicious propaganda but Buchenwald, Auschwitz and Belsen live vividly enough in the memory to save this generation from any offensive sense of superiority to its ancestors. Nor do the still more recent Chinese accusations that American airmen, in 1952, showered the countryside around Kan-Nan Hsien with voles infected with *Pasteurella Pestis*, the bacillus of bubonic plague, suggest that man's infinite capacity for thinking ill of man is in any way on the wane.[32]

The Black Death concentrated this latent fear and hatred of the Jews into one burning grievance which not only demanded vengeance but offered the tempting extra dividend that, if the Jews could only be eliminated, then the plague for which they were responsible might vanish too. There was really only one charge levelled against the Jews; that, by poisoning the wells of Christian communities, they infected the inhabitants with the plague. The Polish historian, Dlugoss, claimed that they also poisoned the air[33] but this view does not seem to have been at all widely shared. Some of the more fanciful reports alleged that the Jews were working under the orders of a conspiratorial network with its headquarters in Toledo; that the poison, in powdered form, was imported in bulk from the Orient, and that the same organisation also occupied itself in forging currencies and murdering Christian children. But these were decorative frills, the attack on the sources of drinking water was the central issue.

The emphasis on this accusation is surprising. With the exception of the Faculty of Medicine at Paris, which suggested that a minor contributory cause of the epidemic might be the pollution of the wells as a result of earthquakes, none of the contemporary experts seem to have tried to link infection with the drinking of tainted water. There were other ways of spreading the plague which must have seemed at least as plausible to medieval man. Alfonso of Cordova's vision of the infection of air by the release of a 'certain confection' into a 'strong, slow wind' has already been mentioned[34] and in subsequent epidemics Jews were accused of passing around clothes taken from the dead or smearing walls and windows with an ointment made from the buboes of plague victims.

A partial explanation may be that many wells in built-up areas were polluted by seepage from nearby sewage pits. The Jews, with their greater understanding of elementary hygiene, preferred to draw their drinking water from open streams, even though these might often be farther from their homes. Such a habit, barely noticed in normal times, would seem intensely suspicious in the event of plague. Why should the Jews shun the wells unless they knew them to be poisoned and how could they have such knowledge unless they had done the poisoning themselves? This theory is supported by Tschudi who, in the *Helvetian Chronicle*, records not only that the Jews knew the wells to be contaminated by 'bad, noxious moistures and vapours' but also that, in many places 'they warned the people against them.'[35] If they did, the warnings seem to have gone unheeded and certainly those who received them were little disposed to feel gratitude to the Jews for their consideration.

There can be little doubt that the majority of those who turned on the Jews believed in the literal truth of the accusations against them. It might be thought that this certainty would have been shaken by the fact that Jews died as fast as Christians; probably faster, indeed, in their crowded and unhealthy ghettoes. But the Christians seem simply to have closed their eyes to reality. Since the Jews caused the Black Death it was ridiculous to suppose that they could also suffer from it. Any appearance to the contrary was merely further evidence of their consummate cunning:

> 'Judée la honnie
> La maulvaise, la desloyal
> Que bien net et aime tout mal
> Qui tant donne d'or et d'argent
> Et promis à crestienne gent
> Que puis, rivières et fonteines
> Qui estoient clares et seines
> En pluseurs lieus empoissonèrent'[36]

But though such crude suspicions might have been acceptable to the mob, they can hardly have been taken seriously by the intelligent and better educated. Dr Guerchberg has analysed the attitude of the leading plague tractators.[37] The most remarkable feature is how few references there are to the guilt or innocence of the Jews. Konrade of Megenberg brusquely dismissed the accusations : 'Some say that this was brought about by the Jewish people, but this point of view is untenable.' In his *Buch der Natur* he cites as evidence Jewish mortality in Vienna which was so high that a new cemetery had to be constructed. Gui de Chauliac was equally categoric. Alfonso of Cordova considered that, by all the rules of planetary action, the Black Death should only have lasted a year and that any subsequent extension must be the result of a wicked plot. But he did not specifically accuse the Jews of being responsible. The 'Five Strasbourg Physicians' warned against poisoned food and water but it is doubtful whether they believed that the poisoning was done deliberately by man.[38] No other tractator paid any attention to the possibility that some human agency was involved in the spread of the plague, still less that such villains must be identified as the Jews.

On the whole, this reticence on the part of the tractators must be taken to indicate that they did not believe the accusations. It is impossible that they did not know what had been suggested and, if they had really thought that a principal cause of the plague was the poisoning of the wells by Jews, then they could hardly have failed to say so in their examination of the subject. Their silence might imply that they thought the idea too ridiculous to mention but it is more likely that they shrank from expressing publicly an unpopular view on an issue over which people were dangerously disturbed.

For it took considerable moral courage to stand up for the Jews in 1348 and 1349 and not many people were prepared to take the risk. The first cases of persecution seem to have taken place in the South of France in the spring of 1348, and, in May, there was a massacre in Provence. Narbonne and Carcassone exterminated their communities with especial thoroughness. But it is possible that the madness might never have spread across Europe if it had not been for the trial at Chillon in September 1348 of Jews said to have poisoned certain wells at Neustadt and the disastrous confessions of guilt which torture tore from the accused.[40] Balavignus, a Jewish physician, was the first to be racked. 'After much hesitation,' he confessed that the Rabbi Jacob of Toledo had sent him, by hand of a Jewish boy, a leather pouch filled with red and black powder and concealed in the mummy of an egg. This powder he was ordered, on pain of excommunication, to throw into the larger wells of Thonon. He did so, having previously warned his friends and relations not to drink the water. 'He also declared that none of his community could exculpate themselves from this accusation, as the plot was communicated to all and all were guilty of the above charges.' Odd scraps of 'evidence' were produced, such as a rag found in a well in which it was alleged that the powder, composed largely of ground-up portions of a basilisk, had been concealed. Ten similar confessions were racked from other unfortunates and the resulting dossier sent to neighbouring cities for their information and appropriate action.

So incriminating a confession settled the doubts or perhaps quietened the consciences of many who might otherwise have felt bound to protect the Jews. On 21 September, 1348 the municipality of Zurich voted never to admit Jews to the city again. In Basle all the Jews were penned up in wooden buildings and burned alive.[40] 'In the month of November began the persecution of the Jews,' wrote a German chronicler.[41] Henry of Diessenhoven has recorded the movement of the fever across his country. In November 1348 the Jews were burnt at Solothurn, Zofingen and Stuttgart; in December at Landsberg, Burren, Memmingen, Lindau; in January, Freiburg, Ulm and Speyer. At Speyer the bodies of the murdered were piled in great wine-casks and sent floating down the Rhine. In February it was the turn of the Jews at Gotha, Eisenach and Dresden; in March, Worms, Baden and Erfurt.

In most cities the massacres took place when the Black Death was already raging but in some places the mere news that the plague was approaching was enough to inflame the populace. On 14 February, 1349, several weeks before the first cases of infection were reported, two thousand Jews were murdered in Strasbourg; the mob tore the clothes from the backs of the victims on their way to execution in the hope of finding gold concealed in the lining. In part at least because of the anti-Semitism of the Bishop, the Jews of Strasbourg seem to have suffered exceptionally harshly. A contemporary chronicle puts the grand total of the slaughtered at sixteen thousand[42]—half this would be more probable but the Jewish colony was one of the largest of Europe and the higher figure is not totally inconceivable.

From March until July, there was a lull in the persecution. Then the massacre was renewed at Frankfurt-am-Main and, in August, spread to Mainz and Cologne. In Mainz, records one chronicler, the Jews took the initiative, attacked the Christians and slew two hundred of them. The Christian revenge was terrible—no less than twelve thousand Jews, 'or thereabouts,' in their turn perished[43]. In the North of Germany, Jewish colonies were relatively small, but their insignificance was no protection when the Black Death kindled the hatred of the Christians. In the spring of 1350, those Jews of the Hansa towns who had escaped burning were walled up alive in their houses and left to die of suffocation or starvation. In some cases they were offered the chance to save

31 December 1349

30 June 1349

31 December 1348

30 June 1348

DENMARK

Dublin Lancaster York Durham

Leicester Norwich Han

London

Bristol

Calais

Amiens Liege Cologne Er

NORMANDY Paris Strasbourg

Angers BA Zuri

Bordeaux Milan V

BEARN Avignon Fl

Montpellier Pisa

CASTILE ARÁGON Marseille

Barcelona

Teruel

Valencia MAJORCA MINORCA

Seville

ANDALUSIA

Almeria

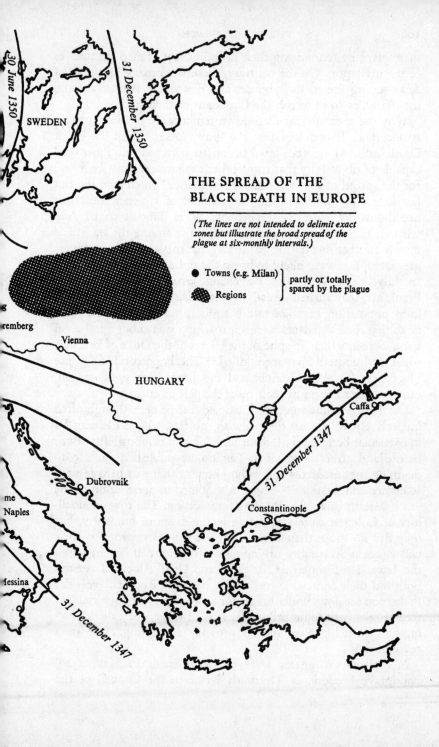

THE SPREAD OF THE
BLACK DEATH IN EUROPE

(The lines are not intended to delimit exact zones but illustrate the broad spread of the plague at six-monthly intervals.)

● Towns (e.g. Milan)   } partly or totally
▦ Regions   spared by the plague

30 June 1350

31 December 1350

SWEDEN

Vienna

HUNGARY

Caffa

Dubrovnik

31 December 1347

Constantinople

Naples

Messina

31 December 1347

remberg

themselves by renouncing their faith but few availed themselves of the invitation. On the contrary, there were many instances of Jews setting fire to their houses and destroying themselves and their families so as to rob the Christians of their prey.

Why the persecutions died down temporarily in March, 1349, is uncertain. It could be that the heavy losses which the Black Death inflicted on the Jews began to convince all those still capable of objectivity that some other explanation must be found for the spread of the infection. If so, their enlightenment did not last long. But the blame for the renewal of violence must rest predominantly with the Flagellants. It is difficult to be sure whether this was the work of a few fanatics among the leaders or merely another illustration of the fact that mass-hysteria, however generated, is always likely to breed the ugliest forms of violence. In July, 1349, when the Flagellants arrived in procession at Frankfurt, they rushed directly to the Jewish quarter and led the local population in wholesale slaughter. At Brussels the mere news that the Flagellants were approaching was enough to set off a massacre in which, in spite of the efforts of the Duke of Brabant, some six hundred Jews were killed.[44] The Pope condemned the Flagellants for their conduct and the Jews, with good reason, came to regard them as their most dangerous enemies.

On the whole the rulers of Europe did their best, though often ineffectively, to protect their Jewish subjects.[45] Pope Clement VI in particular behaved with determination and responsibility. Both before and after the trials at Chillon he published Bulls condemning the massacres and calling on Christians to behave with tolerance and restraint.[46] Those who joined in persecution of the Jews were threatened with excommunication. The town-councillors of Cologne were also active in the cause of humanity, but they did no more than incur a snub when they wrote to their colleagues at Strasbourg urging moderation in their dealings with the Jews. The Emperor Charles IV and Duke Albert of Austria both did their somewhat inadequate best and Ruprecht von der Pfalz took the Jews under his personal protection, though only on receipt of a handsome bribe. His reward was to be called 'Jew-master' by his people and to provoke something close to a revolution.[47]

Not all the magnates were so enlightened. In May, 1349, Landgrave Frederic of Thuringia wrote to the Council of the

City of Nordhausen telling them how he had burnt his Jews for the honour of God and advising them to do the same.[48] He seems to have been unique in wholeheartedly supporting the murderers but other great rulers, while virtuously deploring the excesses of their subjects, could not resist the temptation to extract advantage from what was going on. Charles IV offered the Archbishop of Trier the goods of those Jews in Alsace 'who have already been killed or may still be killed' and gave the Margrave of Brandenburg his choice of the best three Jewish houses in Nuremberg, 'when there is next a massacre of the Jews.'[49] A more irresponsible incitement to violence it would be hard to find.

Nor were those rulers who sought to protect the Jews often in a position to do much about it. The patrician rulers of Strasbourg, when they tried to intervene, were overthrown by a combination of mob and rabble-rousing Bishop. The town-council of Erfurt did little better while the city fathers of Trier, when they offered the Jews the chance to return to the city, warned them quite frankly that they could not guarantee their lives or property in case of further rioting. Only Casimir of Poland, said to have been under the influence of his Jewish mistress Esther, seems to have been completely successful in preventing persecution.

An illustration of the good will of the rulers and the limitations on their effective power comes from Spain. Pedro IV of Aragon had a high opinion of his Jewish subjects. He was therefore outraged when the inhabitants of Barcelona, demoralised by the Black Death and deprived, through the high mortality and the flight from the city of the nobles and the rich, of almost any kind of civic authority, turned on the Jews and sacked the ghetto. On 22 May, 1348, he sent a new Governor to the city and gave orders that the guilty were to be punished and no further incidents allowed.[50] A week later he circularised his authorities throughout the Kingdom ordering them to protect the Jews and prevent disturbances.[51] By February, 1349, the new Governor of Barcelona had made no progress in his search for those responsible. King Pedro grew impatient and demanded immediate action. In a flurry of zeal a few arrests were made, including Bernal Ferrer, a public hangman. But the prosecution in its turn was extremely dilatory. Six months later no judgement had been passed and, in the end, it seems that Ferrer and the other prisoners were quietly released.

Meanwhile, in spite of the King's injunctions, anti-Jewish rioting went on in other cities of Aragon. There was a particularly ugly incident in Tarragona where more than three hundred Jews were killed. Here again Pedro demanded vengeance and sent a commission to investigate. The resulting welter of accusation and counter accusation became so embittered that virtual civil war ensued. In the end this prosecution too was tacitly abandoned. But the King did at least ensure that a new ghetto was built and intervened personally on behalf of several leading Jews who had been ruined by the loss of their houses and documents. When the next epidemic came in 1361 the Jews appealed to the King for protection and an armed guard was placed at the gates of the ghetto.

Flanders was bitten by the bug at about the same time as the Bavarian towns. 'Anno domini 1349 sloeg men de Joden dood'[52] is the chronicler's brutally laconic reference to massacres that seem to have been on a scale as hideous as those in Germany. In England there were said to be isolated prosecutions of Jews on suspicion of spreading the plague but no serious persecution took place. It would be pleasant to attribute this to superior humanity and good-sense. The substantial reason, however, was rather less honourable. In 1290, King Edward I had expelled the Jews from England. Such few as remained had little money and were too unobtrusive to present a tempting target. Some small credit is due for leaving them in peace but certainly it cannot be held up as a particularly shining example of racial tolerance.

The persecution of the Jews waned with the Black Death itself; by 1351 all was over. Save for the horrific circumstances of the plague which provided the incentive and the background, there was nothing unique about the massacres. The Jews had already learned to expect hatred and suspicion and the lesson was not one which they were to have much opportunity to forget. But the massacre was exceptional in its extent and in its ferocity; in both, indeed, it probably had no equal until the twentieth century set new standards for man's inhumanity to man. Coupled with the losses caused by the Black Death itself, it virtually wiped out the Jewish communities in large areas of Europe. In all, sixty large and one hundred and fifty smaller communities are believed to have been exterminated and three hundred and fifty massacres of various dimensions took place. It led to permanent shifts of population, some of which, such as the concentration of Jews in Poland and Lithuania, have survived almost to the present day. It is a curious and somewhat humiliating reflection on human nature that the European, overwhelmed by what was probably the greatest natural calamity ever to strike his continent, reacted by seeking to rival the cruelty of nature in the hideousness of his own man-made atrocities.

SIX

The Rest of
Continental Europe

It would be tedious and probably unprofitable to trace the Black Death in any detail through the remaining countries of Europe. For one thing, the great majority of the more important contemporary chroniclers lived in Italy, France, Germany or England and most of the significant research has been done in those countries. For another, the remarkably similar course which the Black Death took in each country that it visited makes extensive treatment unnecessary. One of the most striking features of the plague is the way in which its principal phenomena are constantly reproduced. The same phrases are used to describe the appearance of the disease, the same exaggerated estimates of mortality appear, the same passions are aroused, the same economic and social consequences ensue.

The similarity is to some extent illusory. The impression is derived partly from the common Latin tongue, used by almost all the chroniclers, and partly from the authority of the Catholic Church which imposed an appearance of uniformity on widely differing social formations. Beneath this veneer of uniformity there were, of course, important variations of wealth, culture and national temperament, but the bracket within which such variations could operate, though to-day being rapidly narrowed again by technological advances, was more restricted in the Middle Ages than at any subsequent period. Any generalisations about Europe as a unit must be perilous, but they will be much less perilous if applied to Europe of the fourteenth century than of the eighteenth or nineteenth. The impact of the Black Death was broadly the same in every country which it visited and anyone observing its effects in Italy and France could have predicted, with a fair degree of success, what would happen as it ravaged Germany and England.

Odd details stand out from the records, often of little more than anecdotal interest. 'During the second year,' recorded the Greek historian Nicephoros Gregoras[1] who was himself a witness of the plague in Constantinople, 'it invaded the Aegean islands. Then it attacked the Rhodians as well as the Cypriots and those colonising the other islands . . . The calamity did not destroy men only but many animals living with and domesticated by men. I speak of dogs and horses, and all the species of birds, even the rats that happened to live within the walls of the houses . . .'

'The second year,' from the context, must mean 1348. Other evidence suggests that Cyprus, at least, was attacked in the late summer of 1347. It seems to have been afflicted with unfair harshness. While the plague was just beginning a particularly severe earthquake came to complete the work of destruction. A tidal wave swept over large parts of the island, entirely destroying the fishing fleets and olive groves on which the prosperity of the Cypriots largely depended. The islanders massacred their Arab slaves, for fear that these should somehow take advantage of the disturbances to get the upper hand, and fled inland. But flight seems to have availed them little. '. . . A pestiferous wind spread so poisonous an odour that many, being overpowered by it, fell down suddenly and expired in dreadful agonies.' 'This phenomenon,' exclaimed the German historian Hecker in justified surprise, 'is one of the rarest that has ever been observed.'[2]

Dalmatia seems to have received the plague across the Adriatic from North Italy. Dubrovnik was attacked in January, 1348, and Split about two months later. In this latter city the wolves, unlike their more superstitious colleagues in Styria, saw the Black Death as nothing but a happy accident which immeasurably improved the prospects for the season's hunting. They 'came down from the mountains and fell upon the plague-stricken city and boldly attacked the survivors.'[3] The rate of mortality was so high that the authorities gave up any pretence of trying to cope and left the dead piled in the streets for weeks at a time.

Scandinavia was attacked by way of England. According to Lagerbring[4] it was carried by one of the wool ships which sailed from London in May, 1349. A member of the crew must have caught the plague just before sailing. The symptoms developed when the ship was at sea and the disease spread so rapidly that within a few days all the crew were dead. The vessel drifted help-lessly until at last it ran aground somewhere near Bergen. The perplexed Norwegians ventured aboard and discovered, too late, what sort of cargo their visitors had brought. The story is pic-turesque and could well be true, though it is sure that Norway would sooner or later have been infected by some other, if less macabre means.

In 1350, King Magnus II of Sweden took somewhat belated alarm. He addressed a letter to his people, saying: 'God for the sins of men has struck the world with this great punishment of sudden death. By it most of the people in the land to the west of our country are dead. It is now ravaging Norway and Holland and is approaching our kingdom of Sweden.'[5] He ordered the Swedes to abstain on Friday from all except bread and water, to walk with bare feet to their parish churches and to process around the cemeteries carrying holy relics. Such measures did as little as else-where to appease the divine anger. Sweden suffered like the rest of Europe and two brothers of the King, Hacon and Knut, were among the victims.

When the Black Death first reached Bergen, several of the lead-ing families and the inhabitants of the chapter-house fled to Tusededal, in the mountains, and began to build themselves a town where they thought that they would be safe.[6] Needless to say, the plague pursued them and carried off the entire community with the exception of one girl. Years later the girl was discovered, still living in the area but run wild and shunning human company. She was christened 'Rype,' meaning wild bird, but seems to have been tamed without undue difficulty, returned to society and married happily. All the land which had been marked out for the new community became the property of herself and her heirs and the 'Rype family' were for several centuries among the large land-owners of the neighbourhood.

The Black Death, or at least its effects, spread north. For many years the Danes and Norwegians had maintained small settle-ments for hunting and fishing in Greenland. Preoccupied by their troubles they now gave up the expeditions and abandoned to their fate any unfortunates who might still have been established there. 'Towering icebergs formed at the same time on the coast of East Greenland,' wrote Hecker in 1832, 'in consequence of the general concussion of the earth's organism; and no mortal, from that time forward, has ever seen that shore or its inhabitants.'[7]

Spain deserves rather less cursory treatment. No major study of the Black Death in the Iberian Peninsula has yet been written[8] but some valuable pioneer work has recently been undertaken, notably by Dr A. Lopez de Meneses,[9] and it is now possible to piece together a rough picture of what happened.

It must first be said that Spain, in the middle of the fourteenth century, was physically and psychologically quite as ill-equipped to face the plague as either France or Italy. Moorish Granada to the south was racked by dissension and depressed by the recent defeats it had suffered at the hands of King Alfonso XI of Castile. But Castile itself was little better. Alfonso, by his triumphal conquest of Algeciras, had the prestige of victory to bolster his position. The preparations which he was making for a siege of Gibraltar promised still greater glory. But the price which his country had paid was exhaustion and impoverishment; high enough when one remembers the scars which it still bore as a memorial of the civil wars during the infancy of its King. Finally, in Aragon, Pedro IV was grappling with armed rebellion and victory was not won until 1348. Instability, insecurity, the constant dread of destruction in campaigns embarked on by irresponsible rulers bent upon little save their own personal glory and aggrandisement: these were the hall-marks of medieval Spain.

According to Pedro Carbonell, archivist of the Court of Aragon, the Black Death began in the city of Teruel and spread out from there through all Aragon, reaching Saragossa in September or October, 1348. This seems most unlikely. Teruel is far from the sea and all the evidence suggests that Spain was first infected through its ports. The timing too is wrong since the first cases of the plague in Spain were recorded in April and May, 1348, almost six months before it spread to Saragossa.

Carbonell must have known that the Black Death had much earlier reached his master's possession of Majorca. In April, 1348, Pedro IV instructed the Government of Majorca to take steps to prevent the further propagation of the disease.[10] This not very helpful instruction does not seem to have been observed with any success. One chronicler puts the death roll at fifteen thousand in a single month and other writers suggest that the total loss was at least twice that figure and up to eighty percent of the total population.[11] On 3 May, the Government of Majorca were complaining that the island was so weakened by disease that it could no longer protect itself against the attacks of pirates and the Bey of Tunis. They appealed for help from the mainland. Pedro IV agreed to send some galleys but, with the characteristic pawkiness of a central government, insisted that Majorca pay half the cost. By June, 1349, the Governor of Majorca found himself in his turn instructed to send troops to defend the still worse depopulated Minorca against probable enemy attack.

The Mediterranean coast of Spain was quickly affected. Barcelona and Valencia were both struck by the plague in May, 1348, and Almeria in June. Ibn Khātimah, the Arab physician, who was living in Almeria at the time, says that the first case arose in a house in the poor quarter of the town belonging to a family called Beni Danna.[12] The plague took an unusually protracted course, lasting through the autumn and winter and still being active when Ibn Khātimah wrote his record in February, 1349. Even at its peak it was not killing more than seventy people a day in a population that must have been in the neighbourhood of twenty thousand.

The disease spread next through Arab Spain so that the armies confronting Alfonso XI were afflicted before their Christian enemies. It is said that the Arabs were deeply disturbed by this phenomenon and many of them seriously thought of adopting Christianity as a form of preventive medicine. Fortunately for their faith, however, the Black Death was soon raging quite as disastrously among the troops of Castile. 'When they learned that the pestilence had now reached Christian men their good intentions died and they returned to their vomit.'[13] The Castilian army in front of Gibraltar survived inviolate through 1349; then, in March, 1350, was suddenly attacked by the plague. The senior officers begged King Alfonso to leave his troops and seek safety in isolation but he refused to do so. He duly caught the disease and died on Good Friday, 26 March, 1350. He was the only ruling monarch of Europe to perish during the Black Death.

But the royal house of Aragon did not escape unscathed. King Pedro lost his youngest daughter and his niece in May, and his wife in October. By the autumn order seemed to be breaking down in his dominions. Bands of armed brigands were straying over the countryside and an ordinance had to be published ordering severe punishment for any one found looting the houses of plague victims. Li Muisis records the experiences of a pilgrim to St James of Compostella who passed through Salvatierra on the way home.[14] The town had suffered so grievously from the plague that not one citizen in ten remained alive (a characteristically woolly estimate which, *inter alia*, took no account of those people who had fled the town, to return when the plague was past). The pilgrim put up at an inn and 'after taking supper with the host (who, with his two daughters and a single servant was the only survivor of his entire family and who had no sensation of being sick himself), paid for his night's lodging in advance, as he meant to leave at dawn the next day, and went to bed.' Next morning, when he got up, he found that, after all, he had to see the proprietor of the inn. After trying for some time to get an answer he eventually routed out an old woman sleeping in another part of the inn. It was then that he discovered that host, daughters and servant had all died during the night. 'On hearing this,' Li Muisis records, 'the pilgrim made all haste to leave the place.' In his account of the plague Ibn Khātimah referred to its intense infectiousness and to the coughing of blood by the victims; anecdotes of this kind confirm that the pulmonary and septicaemic variants of the Black Death must have been rife in Spain.

Until quite recently it was accepted tradition that the plague scarcely penetrated to Castile, Galicia and Portugal. This is clearly not true, though in general the Atlantic coast of Spain was less severely afflicted than the Mediterranean. The rich of Castile seem to have been peculiarly affected by the urge to give their lands and possessions to the Church in the hope that they might thereby earn themselves freedom from the plague or, at the worst, the guarantee of a comfortable niche in paradise. So far, indeed, did this process go that, when the panic abated, it was found that the economic structure of the country had been dislocated and an altogether undesirable amount of wealth accumulated in the hands of the Church. To redress the balance King Pedro I in 1351 ordered that, where the donors themselves or their heirs could be traced, then the Church must disgorge its gains.

Though Portugal as a whole does not seem to have suffered particularly seriously, the city of Coimbra was devastated. It is

said that ninety percent of the population died—a statistic which
need not be believed. But it seems certain that the prior and all
the prebendaries of the great collegiate monastery of St Peter of
Coimbra perished within a few days, an impressively clean sweep
suggesting that the popular tradition may not have been totally
devoid of substance.

Dr Carpentier has prepared a map of Europe at the time of the
Black Death (reproduced on pp. 104 and 105) showing the
movements and incidence of the plague. Virtually nowhere was
left inviolate. Certain areas escaped lightly: Bohemia; large areas
of Poland; a mysterious pocket between France, Germany and the
Low Countries; tracts of the Pyrenees. Certain others were
afflicted with especial violence: cities mainly—Florence, Vienna,
Avignon—but also whole areas such as Tuscany. A host of factors, some of them still unidentified, played their part in deciding
whether any given area should suffer lightly or severely. The inclinations of the rats must have been the most important: a
shortage of food in one place driving them on, the resistance of
the indigenous rats holding them at bay in another. Climate was
certainly significant; it seems that the bacillus of pulmonary
plague finds it hard to survive in cold weather. The chance movement of an infected human could sometimes save or condemn a
village. Did some people also enjoy a built-in resistance to
bubonic plague? Even to-day the science of epidemiology cannot
provide a fully conclusive answer—the problem of where and
when a disease will strike next is still unsolved.

But such gradations in horror were anyway of minor significance. Though the density of corpses might vary, the smell of
death was over the whole of Europe. Scarcely a village was untouched, scarcely a family did not mourn the loss of one at least
of its members. As the shadow of the Black Death passed away it
must have seemed to those who survived that recovery could
never be a possibility.

SEVEN

Arrival in England:
the West Country

The England of 1348, politically and economically, was not in so frail a state as some of the countries on the mainland of Europe. Indeed, viewed from France, it must have seemed depressingly prosperous and stable. Since Edward III had routed Queen Isabella and the Mortimers at the end of 1330 he had bestridden the narrow world of England, if not like a Colossus, then at least as a figure considerably larger than life. He managed to combine the charismatic appeal of a *beau chevalier sans peur et sans reproche* with the ruthlessness and lack of scruple which every medieval monarch needed if he were to enjoy a reasonable tenure of his throne. His main vice, one not immediately apparent to his subjects, was his stupidity; his second was ambition, spiced with vanity, which drove him on to establish himself as a figure of glory on the international stage. He gave England a unity and a sense of security which it had not enjoyed since the days of his grandfather. But by his determination to have his way, not only in his own country but in Scotland and France as well, he made certain that the profit which England should have gained from its stability was dissipated frivolously on foreign soil.

A conscientious chauvinist could put forward a reasonably good case for maintaining that Edward's wars against France and Scotland were the result of intolerable provocation and conducted strictly in defence of just national interests. A student of politics might maintain that only by foreign victories could Edward III have hoped to win the respect of his nobles and unite his country. For the purposes of this book it is enough to note that, though the French and Scots might be defeated militarily, the English never had the strength fully to follow up their victories in the face of even minimal determination on the part of the enemy. Nor, though temporary truces supervened, did

Edward have either the will or the wisdom to disengage from his campaigns at an advantageous moment and settle for something short of total victory.

At Dupplin Moor, Halidon Hill and Neville's Cross, the English had won spectacular triumphs against the Scots; at Sluys, Crécy and Calais against the French. By 1348 the reputation of English arms can rarely have stood higher and the King's prestige was at its zenith. But the glory was meretricious and the cost in money and man-power mounted steadily. Edward's crown was pawned to the Archbishop of Trier, his debts to the Bardi and Peruzzi were enormous, the rich merchants and the City of London had been repeatedly mulcted, taxes had been raised as high, perhaps even higher than was prudent. England was still a rich country but it was under severe financial strain and the strain was beginning to tell.

It is important not to exaggerate the progress of decay. By and large England was a thriving country. Exports of wool, by far the most important single crop, were buoyant and exports of cloth, a new and rapidly expanding trade, had, by 1347, reached a level sufficiently high to lead the King to impose an export tax.[1] The great territorial magnates, or at least the Dukes and Earls, possessed imposing riches. The Duke of Lancaster, from his English lands alone, had an income of £12,000 a year (translation into modern currency must be hedged around with so many qualifications as to be virtually meaningless but a very approximate order of magnitude can be obtained if one multiplies the medieval money by between fifty and sixty). The Bohuns had £3,000 a year. The Earl of Arundel left a hundred thousand marks in ready money.[2] (The mark could signify several things but thirteen and fourpence is the most usual meaning.) A leading merchant like William de la Pole could lend the King more than £110,000 in a little over a year—not, of course, from his own personal resources but from funds on which he could freely draw.[3]

The wheat-growing and sheep-raising country of East Anglia and the Southern Midlands provided many lesser but comfortable fortunes. And it was not only merchants and gentlemen who were doing well out of the national prosperity. The villein, too, might sometimes enjoy substantial wealth and employ several labourers to help him in his farming. In theory he could own no land and had to be entirely at the beck and call of his master but, in practice, his labour services had often been exchanged for a money payment. In virtually every town a charter of liberties had been granted to its citizens; only in a few cases, usually where the lord was a tenaciously conservative churchman, did the traditional relationship between lord and urban tenant survive unmodified.

Probably a little under twelve percent of the population lived in cities or towns.[4] Of these London was by far the largest and most prosperous. It had eighty-five parishes within its city wall with Westminster, Southwark and other outlying villages closely associated with its daily life. It contained between fourteen and eighteen thousand households, giving it a population that cannot have been far off sixty or seventy thousand.[5] Norwich was almost certainly the second city of the realm with some 13,000 inhabitants and York too may well have had more than 10,000 citizens.[6] After that came a plethora of lesser towns with Winchester, Bristol, Plymouth and Coventry, among the larger.

But the main unit of English life was the village. Indeed, since the isolated dwelling was almost unknown outside the Celtic fringe, it can safely be said that virtually everyone who did not live in a town or city, nearly 90% of the population, was to be found in villages varying in size between the large, of up to 400 inhabitants and the small, which might contain as few as twelve families. Though the anecdotes and the striking statistics will usually have to be culled from the towns and great monasteries, the story of the Black Death in England is above all the story of its impact on the village community. It is in the society of the villages that its most long-lasting results were to be recorded.

'In this year 1348, in Melcombe, in the county of Dorset, a little before the Feast of St John the Baptist, two ships, one of them from Bristol, came alongside. One of the sailors had brought with him from Gascony the seeds of the terrible pestilence and, through him, the men of that town of Melcombe were the first in England to be infected.'[7]

Other ports have been put forward for the doubtful honour of being the first in England to receive the plague. One chronicler, from the Abbey of Meaux in Yorkshire, believed that Bristol was infected earlier,[8] Henry Knighton opted for Southampton,[9] while

John Capgrave, writing some eighty years later, recorded: 'First it began in the north cuntre; than in the south; and so forth thorw oute the reme.'[10] The latter thesis, at least, can be dismissed. The Black Death may well have made a separate entry into the north of England but certainly not until several months later than in the south. Indeed, there is no evidence that the plague took a firm grip in the northern counties until the beginning of 1349.

Many ports of southern England were in constant, almost daily contact with the continent or with the Channel Islands. It would be surprising if trading ships had not carried the plague to several of them. But the consensus of opinion seems to be that Melcombe Regis, now part of Weymouth, earned its unsavoury claim to fame. As well as the *Grey Friars' Chronicle* quoted above, a monk of Malmesbury also refers to 'a port called Melcombe, in Dorsetshire,'[11] and further chroniclers, no doubt in some cases copying the opinions of their contemporaries, either refer to it by name or state that the plague arrived at a 'Dorsetshire port' with other details which fit its description.[12]

The confusion is a great deal worse when it comes to deciding on exactly what date the plague was first observed in England. The monk of Grey Friars states that it arrived 'a little before the Feast of St John the Baptist'—that is to say before 24 June, 1348. *Higden's Polychronicon* also agrees that this was the crucial date. But nobody else is prepared to put it so early. Robert of Avesbury says that the plague began 'about St Peter's Day,'[13] presumably meaning 29 June, rather than the other dates on which the Apostle is commemorated. The monk of Malmesbury opted for 7 July. The Canon of Bridlington favoured 'the feast of St James,' or 25 July. Henry Knighton of Leicester referred to 'the autumn' of the year 1348; an imprecise period but one which could hardly have begun before the end of August. The Bishop of Bath and Wells, on 17 August, ordered 'processional stations every Friday . . . to beg God to protect the people from the pestilence which had come from the East into the neighbouring kingdom.' The reference to the plague in 'the neighbouring kingdom,' which can only mean France, seems to imply that the Bishop was not yet aware that the disease was already to be found in England. Yet it seems incredible that he should not have known about an epidemic which, according to others, had already been raging for nearly two months in his own diocese. Finally, Stephen Birchington deferred the outbreak to immediately after Christmas, 1348.[14] Since, however, he reported that it ended in May, 1349, it is reasonable to detect some confusion in his mind between the duration of the plague in Canterbury and in England as a whole.

This evidence is cited at somewhat tedious length not because it matters much whether the Black Death arrived in Melcombe Regis or in Southampton, a few weeks earlier or a few weeks later, but to illustrate the extraordinary difficulty of establishing with any precision the details of what took place. If the chronicles are unable to agree within three or four months on the date on which the first case of the plague was recorded, then how much more likely it is that there will be complete confusion when such complex problems as the number of thousands slain by the disease come to be discussed. Piecing together the various accounts, the most likely picture of what actually happened is that a ship bearing a victim of bubonic plague did arrive at Melcombe Regis at the end of June, 1348; that the first case of a local inhabitant catching the disease occurred in early July and that the disease did not begin to spread or to develop its terrifying pulmonary and septicaemic variations until the beginning or middle of August. But that, as is the case with so much that will follow, is a guess and the truth will probably never be established with certainty.

If Melcombe Regis was indeed the first port to receive the Black Death it may have been brought from Calais. Melcombe was at this time a town of some importance contributing almost as many ships to the siege of Calais as Bristol or even London. It could well have been one of these ships returning from France which brought in the plague. *Prima facie* the suggestion of the *Grey Friars' Chronicle* that the plague was imported from Gascony is less likely since Melcombe is not known to have received many boats coming from that region. But it is by no means impossible; one of the ships was said by the *Chronicle* to have had its home in Bristol and this could well have been on the return journey from some Gascon port. Another, and perhaps still more probable source of infection, is the Channel Islands. Jersey and Guernsey were suffering badly from the Black Death at this time, so much so that Edward III wrote to the Governor of Jersey:[15]

'By reason of the mortality among the people and fishing folk of these islands, which here as elsewhere has been so great, our rent for the fishing which has been yearly paid us, cannot be now obtained without the impoverishing and excessive oppression of those fishermen still left.'

The letter is undated and it is not known by how far or, indeed, if the Black Death in the Channel Islands preceded the outbreak on the English mainland. But if, as seems probable, the islands were affected first, it is to Melcombe Regis more than to any other English port that they are likely to have spread the disease.

But whether by way of Southampton or by Melcombe Regis; whether in June, July or August; it was inevitable that the Black Death would sooner or later spread to the British Isles. It is tempting to think of Britain isolated behind her sea defences, remote from Europe and, with a bit of luck, immune from the misfortunes of her continental neighbours. But the truth, then as now, is that England was part of the continent of Europe and that the Channel as much linked England and France as divided them. Indeed, it was a great deal easier for men and merchandise to arrive by sea in England than to make the perilous crossings of the Alps or venture along the other land routes of continental Europe. 'The south-east of England,' wrote Professor Kosminsky,[16] 'lay at a great cross-roads where the trade routes from Scandinavia, the Baltic, the North Sea, the Atlantic coast and the Mediterranean all met, as well as the great river-ways of the Rhine, the Meuse, the Scheldt and the Seine.' Along the trade routes, without possibility of check, moved the Black Death.

From Melcombe Regis the plague struck inland across the West Country. It is not difficult to get an approximate idea of its course. England is badly endowed with the impressionistic reporting of such chroniclers as Michael of Piazza or Agnolo di Tura. There is even less in the way of dispassionate medical records; English physicians contributed virtually nothing to the ample if somewhat profitless literature of the plague tractators. But the richness of our national archives—the archives of a nation wedded to legalism and the virtues of precedent and, still more important, of a nation which has had the good fortune never to suffer subsequent foreign invasion—offers a fuller picture of the progress of the Black Death than those which any other country can provide.

Professor Stengers, the Belgian historian, referred wistfully to the riches of the English archives as being the envy of every continental medievalist.[17] Envy certainly; and yet it would be surprising if the continental historian did not sometimes feel a certain relief and the proud possessor of this treasure-house occasionally view his national glory with apprehension as well as pride. The knowledge that untapped reservoirs of knowledge exist, ready to confound the over-confident and ensnare the unwary, is sobering even to the expert and downright intimidating to the amateur.

This seems particularly true when material about the effects of the Black Death is in question. The most complete source, though by no means the most comprehensive, is the ecclesiastical records. Cardinal Gasquet relied almost exclusively on these for his study of the Black Death. The principal series which he used were the Books of Institutions, showing dates of appointment to the various livings, together with the Patent Rolls which listed, *inter alia*, 'royal grants, licences and presentations made by the Sovereign to such vacant ecclesiastical livings as were at the time in the royal gift.'[18] The value of such records is obvious but, as will be seen later, so also are their limitations. For the moment it suffices to say that, as a rough guide to the date that the Black Death was raging in any particular area and to the relative damage which it did in one place or another, better evidence is rarely to be found.

Among the lay documents, those which are of immediate relevance for a study of the Black Death are the manorial Court Rolls. From these, in ideal circumstances, it is possible to establish how many householders died in any given period and whether there were relations left to inherit or the holding escheated to the Lord. Though once again such lists pose problems when it comes to deducing from them a comprehensive total of plague victims they are of the utmost value in that they show the incidence of the Black Death in each manor. Read in conjunction with the Account Rolls they provide an amazingly detailed picture of life on the medieval manor. But many fewer of them are left than is the case with the ecclesiastical documents and, as a general rule, they also tend to be less well kept and less accessible. A series is necessary to enable valuable deductions to be drawn—yet too often the series is interrupted and only isolated numbers survive.

With the help of the ecclesiastical records, it can be established that the plague was rife in many parts of Dorset by October, 1348, reached a peak in December and January and was on the wane by the end of February. New vicars had to be appointed at Shaftesbury on 29 November, 10 December, 6 January and 12 May and Wareham lost the head of its Priory in October and had two new vicars instituted in December, one in May and another in June.

Exactly one hundred institutions to Dorset benefices caused by the death of the previous incumbent were made during the seven months from October, 1348 to April, 1349. The numbers did not return to normal until the autumn of 1349.[19] From other sources one learns that Poole was particularly badly affected and that a tongue of land projecting into the sea and known as 'The Baiter' was bought by the town-councillors and set aside as a burial place for the victims. At Bridport, though the plague was not so bad as to interfere with the supply of cordage to the royal navy,[20] two additional bailiffs had to be appointed to cope with the extra work.

In January, 1349, Ralph of Shrewsbury, Bishop of Bath and Wells, circulated a letter to all the priests in his diocese which shows up vividly the demoralisation in the infected areas:[21]

'The contagious pestilence of the present day, which is spreading far and wide, has left many parish churches and other livings in our diocese without parson or priest to care for their parishioners. Since no priests can be found who are willing, whether out of zeal and devotion or in exchange for a stipend, to take on the pastoral care of these aforesaid places, nor to visit the sick and administer to them the Sacraments of the Church (perhaps for fear of infection and contagion), we understand that many people are dying without the Sacrament of Penance. These people have no idea what recourses are open to them in such a case of need and believe that, whatever the straits they may be in, no confession of their sins is useful or meritorious unless it is made to a duly ordained priest. We, therefore, wishing, as is our duty, to provide for the salvation of souls and to bring back from their paths of error those who have wandered, do strictly enjoin and command, on the oath of obedience that you have sworn to us, you, the rectors, vicars and parish priests in all your churches, and you, the deans elsewhere in your deaneries where the comfort of a priest is denied the people, that, either yourselves or through some other person you should at once publicly command and persuade all men, in particular those who are now sick or should fall sick in the future, that, if they are on the point of death and can not secure the services of a priest, then they should make confession to each other, as is permitted in the teaching of the Apostles, whether to a layman or, if no man is present, then even to a woman. We urge you, by these present letters, in the bowels of Jesus Christ, to do this . . . And, in case anyone might fear that a lay confessor would make public the confessions which they heard and, for this reason, might hesitate to confess himself to such a person even in time of need, you should announce to all in general and, in particular, to those who might hear confessions in this way, that they are bound by the laws of the Church to conceal and keep secret such confessions and that they are prohibited by sacred canonical decrees from betraying such confessions by word, sign, or any other means, except at the wish of those who have made such confession. If they break this law then they should know that they commit a most grievous sin and, in so doing, incur the wrath of Almighty God and of the whole Church.'

The Bishop concluded:

'The Sacrament of the Eucharist, when no priest is available, may be administered by a deacon. If, however, there is no priest to administer the Sacrament of Extreme Unction, then, as in other matters, faith must suffice.'

Even with this exception, it is clear that the Bishop was authorising a very considerable relaxation of the normal rules.

The authority to hear confession has, in all periods of the Church's history, been restricted to the priesthood. To throw it open to laymen and even to women, though not in defiance of canonical authority, was a step to be taken only in case of extreme emergency. It was a confession on the part of the Church that the crisis was out of control and the normal machinery no longer able to cope with it.

The most revealing phrase in the Bishop's letter is the one in which he refers to the lack of priests willing to take on new parishes or to visit the sick 'perhaps for fear of infection and contagion.' The implied rebuke would have come better if Ralph of Shrewsbury himself had ventured a little farther into the battle. From November, 1348, until 13 May, 1349, the period in which the Black Death was at its height in all parts of his diocese, the Bishop remained at his house at Wiveliscombe, a remote village in the corner of his territory.[22] It is true that it was his normal practice to winter at Wiveliscombe and also only justice to say that he seems in no way to have neglected his duty or shunned direct contact with visitors from plague-infested areas. Indeed, a stream of priests came to his retreat to receive their letters of institution. No doubt he had good reason to argue, like Pope Clement before him, that the best way he could serve his flock was by staying alive and not indulging in false or, at least, futile heroics. But, when all is said and done, one would still have slightly greater respect for the Bishop and sympathy for his railing at the reluctant clergy if he had paid a single visit to Bristol, Bath or any other important town in his diocese while it was suffering the agonies of the plague.

However reluctant some of the priests may have been to expose themselves, the clergy of Somerset, another county in the Bishop's diocese, did in fact suffer greatly as a result of the Black Death.[23] Institutions to new benefices rose from a more or less normal figure of nine in November, 1348, to thirty-two in December, forty-seven in January, 1349, forty-three in February, thirty-six in March, forty in April and then fell away to twenty-one in May and a mere seven in June—the month in which the Bishop thought fit to set forth on his travels again. So extreme was the confusion that the Bishop felt it necessary to insert a saving clause in all his appointments protecting his position in case, in a moment of excusable aberration, he instituted a priest to a benefice which in fact was not vacant at all. It would be most unwise to generalise on the basis of a single county but it is fair to say that the evidence of Somerset shows no tendency on the part of the parish priests to shirk their terrifyingly perilous responsibilities.

Such data needs closer analysis before they can provide more than an indication of a general trend and often the material for such an analysis does not exist. Though Gasquet himself does not mention the fact it seems, for instance, that in the case of Somerset about a quarter of the new institutions were the result of the resignation of the previous incumbent rather than his death.

But, in its turn, for this figure to mean much one would have to know what inspired the individual resignation. Was it reluctance to face the dangers which confronted a parish priest during a lethal epidemic, the economic impossibility of soldiering on in an anyway poor parish which had now lost the majority of its more prosperous parishioners or, perhaps most probable, the translation of the incumbent to another, more important parish which had lost its priest? Even among those who died the statistics are not wholly conclusive since one or two at least may have been the victims of old age, accident or other disease rather than the Black Death. Such facts will never be established: the historian is lucky even if he finds proof that the vacancy was caused by death, let alone information about its cause.

Professor Hamilton Thompson has pointed out other considerations which throw doubt on statistics of this kind. For one thing, the place of death is rarely specified. If a Yorkshire parish priest died of the Black Death while on duty in Canterbury or a priest from a rural Hampshire parish preferred to tend his flock from his comfortable house in Winchester, then it would be misleading to quote his death as evidence of the mortality in his proper county or his proper deanery. Another flaw is that a few institutions were not recorded in the Register, presumably because of the muddle and stress caused by the hurried appointment of a quite abnormal number of new priests at a time when the officials responsible were themselves leaving their posts vacant with alarming speed. For certain important areas, too, the records are not available. And finally, it has proved virtually impossible to establish a list of benefices which can categorically be stated to be complete.[24] Calculations made on such a basis still possess great value, but almost always they must be used with caution and a certain scepticism.

A fortiori this is true when figures for the mortality among the clergy are extended to cover laics as well. It would be extremely rash to accept unquestioned Cardinal Gasquet's firm assertion: 'It cannot but be believed that the people generally suffered as greatly as the clergy, and that, proportionally, as many of them fell victims to the scourge.'[25] It can be contested that the beneficed clergy, with their education, higher standard of living and less cramped living conditions, were much better placed to survive than their unfortunate flocks. But, on the whole, the arguments which suggest a higher death rate among priests than laics are more convincing. For one thing there was the nature of their work which, if conscientiously carried out, brought them into constant contact with the infected. In particular in the areas where the pulmonary form of the disease was rife, this must have been close to a sentence of death on any priest resolved to do his duty. For another, as Professor Russell has pointed out, the fact that the average age of the clergy was higher than that of the population as a whole meant that, in any given year, a higher proportion of priests were likely to die.[26] And finally, though the smaller size of the priestly household reduced the chances of infection, it seems also to have been the case that, once such a household was infected, the chances of any survivals were proportionately less. One rat family to a household and three fleas to a rat seems to have been the norm; the greater the number of infected fleas in proportion to potential human victims, the smaller the chances of escape.

No final answer to this conundrum will ever be forthcoming. But it would be reasonable to state as a general rule that the proportion of beneficed clergy who died in any given diocese could not possibly have been much smaller than the corresponding figure for the laity and is unlikely to have been *very much* bigger. Arbitrary limits of 10% less and 25% more seem to provide a reasonable bracket within which the correct figure must be encompassed.

Dr Lunn has calculated that 47.6% of the beneficed clergy in the diocese of the Bishop of Bath and Wells died of the Black Death. It can therefore be said, with reasonable confidence, that it is unlikely that more than 52% or less than 35% of the total population met a similar fate. A safer, because looser way of expressing the same proposition would be to say that, taking a conservative view, between a third and half the people must have died.

The figures for the mortality among beneficed clergy can be used with much greater confidence when it comes to establishing a ratio between different areas. If twice as many clergy died in Yorkshire as in Northamptonshire, then it is reasonable to assume that more or less twice as many laymen died as well. It would be tempting to apply the same mathematics to cities and towns as well as larger areas, but, obviously, the narrower the statistical base, the more risk there is of serious distortions being introduced. It is permissible to compare dioceses on this basis, possibly even archdeaconries, but where deaneries or smaller units are concerned then the comparative figures are no more than a valuable but uncertain pointer towards the relative sufferings of the areas.

In December, 1349, when things were almost back to normal, Bishop Ralph ventured as far as Yeovil. As part of his visitation he held a special service of thanksgiving. To his dismay certain 'sons of perdition' armed with 'bows, arrows, iron bars and other kinds of arms' attacked the church, injured many of his attendants and kept the Bishop and his congregation bottled up until nightfall. The siege was then transferred to the rectory where it lasted till the following day. At this point the sons of perdition either got bored and went home or, as the official story had it, a party of 'devout sons of the church' came to the rescue.[27] Sixty of those concerned were later ordered to do public penance.

It is tempting to read a perhaps impermissible amount into this story. It showed, after all, extreme audacity on the part of the inhabitants of Yeovil to attack a magnate as powerful, both spiritually and temporally, as the Bishop of Bath and Wells. Though the mild revenge which he exacted suggests that the assault was not particularly serious, there must still have been good reason, in the minds of the rioters at least, to indulge in such an escapade. Some particular grievance may have inspired it but the action of the crowd may surely reflect considerable anger against the Bishop and all the ruling classes, a by-product of the intense fear and misery in which they had lived for the previous twelve months. Pressures of that kind must generate intense emotions and such emotions require an outlet.

To appreciate the full impact of so fearful a calamity on an ignorant and credulous people calls for an intense effort of historical imagination. Some glimmering of what it was like might be gleaned from the reactions of the people of London and Coventry or, still more, of Dresden and Berlin in the face of prolonged and devastating air-attack. In these cases a pattern of crowd behaviour has been established. There was an initial reaction of anger directed against the enemy, exhilarating, almost euphoric,

with vows of vengeance and pride in the courage and solidarity
which the victims of the bombing displayed. Then there might be
panic, a brief breakdown of morale and the capacity to produce
disciplined or rational responses to external stimuli. And finally
came apathy and indifference; a grudging though often successful
adaptation of life to the needs of the new situation.

But with apathy came rancour and suspicion; doubts about the
other members of society with whom, so recently, they had felt
united in suffering. Suspicions of the rich: the more prosperous
parts of town, said the poor, were mysteriously spared by the
raiding bomber—this proved that some sinister understanding
existed between them and the enemy. Suspicions of the rulers:
they only kept the war going so as to grow fat on arms sales or for
some other selfish end. Suspicions of the doctors: they saved their
drugs for themselves or their privileged friends. Suspicions of the
shop-keepers: they hoarded their precious goods to sell at a profit
to the undeserving who could afford to pay. Class looked askance
at class; neighbourhood at neighbourhood. Loyalties retracted:
to the street; to the family; ultimately to oneself.

The analogy between twentieth-century air-raid and medieval
pestilence obviously breaks down at many points. Baehrel has
suggested that the Revolutionary Terror in France produced the
same defence reactions in those who endured it as were to be
found in the plague-struck Europeans of 1348 or, for that matter,
the victims of the cholera epidemic of 1884.[28] The conclusions
which he draws are strikingly similar to those derived from a
study of the blitz. A belief in plots; a conviction that someone has
to be made the scapegoat for everything; were almost always the
chosen outlet for surplus passions. Whether aristocratic emigré or
Jewish poisoner, bourgeois tool of a reactionary clique or incompetent doctor, '. . . the suspect came to the fore; a suspect
who was not the same for all. For the well-fed the suspect was the
poor man because he was dedicated to the plague; for the lower
classes the suspects were the rich, in whom they were quick to
identify the propagators of the disease. For some the leading
suspect was the surgeon . . . for others the beggar . . .'

And yet neither the blitz nor the Revolutionary Terror yield
an adequate impression of the psychological shock which medieval
man endured. For where there is a common, identifiable enemy
then there must be a sense of camaraderie; it matters little who

the foe may be, to hate him will provide relief and bulk larger in
the mind than the pettier grudges that divide one from one's
neighbours. The first fine flower of anger against the emigré
aristocrat or the marauding bomber might, in time, lose its capacity
to excite or inspire but it survived as an element lending cohesion
to the attacked. Medieval men had no one to hate. They might
work off their resentment in campaigns against the lepers or the
Jews but few of those who sacked the ghettoes can have believed
that, by their deed, they were doing more than tinker with the
instrument of their destruction while leaving the root cause untouched. The Black Death was the work of God, and against God
they could not fight.

The only defence against the plague in which the doctors had
the slightest faith was flight from the afflicted area. This the poor
knew, and yet they knew too that it was a defence to which they
could have no recourse. As the poor of Genoa, Florence, Paris or
London saw the rich and privileged bundle up their most precious
possessions and flee the cities it would have been astonishing if
they had felt no resentment, no sense that they were being
deserted and betrayed. With such a mood abroad it was inevitable
that the processions of the Flagellants would quickly take on a
revolutionary tinge, that the houses of the magnates would be
sacked and the clergy abused, derided or even assaulted.

There is little chapter and verse to illustrate the upsurge of
class hatred which arose during the plague. 'Before 1789,' wrote
Baehrel in explanation of this in France, 'this sentiment of hatred
left few traces: the poor rarely use a pen.' But subsequent epidemics have made it clear how quickly the feelings of the under-
privileged could be embittered. During the cholera epidemic of
1832, when slightly greater sophistication if not tolerance might
have been expected, the Parisian mob rioted through the smarter
quartiers, accusing nobles and bourgeois not only of suffering less
seriously from the disease but of poisoning their impoverished
fellow-citizens into the bargain. Who can doubt that the vastly
more credulous and worse afflicted poor of the fourteenth
century must have felt the same rancour and suspicion? If they
failed to sack the houses of the rich it can only have been because
the torpor induced by famine and misery had already broken their
spirits before the plague began to work on their emaciated bodies.
But, in the last analysis, the most noticeable feature of the

Black Death was not that some escaped but that everyone was to some extent involved and paid the price of involvement. For the months which the Black Death lasted it must have seemed to those who suffered that everything was discredited and at an end. The doctors could cure nobody and, by their efforts, made themselves a laughing stock. The Church was impotent to defend itself or its faithful and had resort only to muttered objurgations about the sinfulness of mankind. The rulers abandoned their palaces and their responsibilities and left their people to die in misery. And the Black Death spared nobody.

> 'Sceptre and crown
> Must tumble down
> And in the dust be equal made
> With the poor crooked scythe and spade.'

Death had always been a preoccupation of medieval man; now it became an obsession. Always he had known that in time it must come to everyone but never before had the fact been brought so forcibly to his attention. Never before had those set in authority over him been shown so clearly to be no braver, no better, no wiser and no less vulnerable. Like every other lesson, it was to be forgotten but, at that moment, it must have seemed that its memory would never fade.

It is impossible that England should have been spared such tensions but even the somewhat scanty evidence for their existence which is to be found in the countries of continental Europe is lacking this side of the Channel. The maltreatment of the Bishop of Bath and Wells, which was mentioned at the beginning of this digression, could possibly have had such an origin, yet equally some quite different factors, of which we now know nothing, may have been responsible. A monk was beaten up in Winchester yet, as we shall see, there was good and sensible reason for his misfortune.[29] A spirited battle between monks and townsmen took place in Hull but such affrays, in Hull, were practically a local sport and call for no special explanation. The excesses of the Flagellants found no favour with the people of London and the few Jews who still lived in England were left in peace. The Bishops were constantly at work to whip up penitential fervour and not to curb it. A few incidents of panic or violence can be culled from the contemporary chronicles but nothing remotely suggestive of mass-hysteria.

Can one deduce from this that the Englishman, in the face of quite as grave a danger, proved more phlegmatic or better disciplined than his continental contemporary? It would be hazardous to push the argument too far. To argue that something must be true because of lack of evidence to the contrary is always dubious. When the evidence either way is as scanty as in England of the fourteenth century it would be folly. But what can be said with fair confidence is that any widespread movements on the scale of those experienced in Spain, France or Germany could not have escaped the attention of the chronicler. For one reason or another the Englishman did not indulge in the massive disorders in which others found an outlet for their emotions.

There is no reason to exclude national temperament from the complex of factors which must explain this omission provided that one does not try to erect too pretentious or elaborate a structure on the small basis of established fact. Even in the fourteenth century, when inadequate communications and the weakness of the central government ensured that loyalties were still primarily to the lord, the community or the region, there was already apparent a consistency in English life and character which it would be absurd altogether to ignore.

'They could not, they would not be driven or frightened out of what they dimly comprehended they had to do.' The words were applied by Drew Middleton to the Londoner in the blitz[30] but they fit as well in the fourteenth century. One of the most striking features of the Black Death in England, attested to in the Court Rolls of innumerable manors and those borough records that are still available, is the way in which communal life survived. With his friends and relations dying in droves around him, with labour lacking to till the fields and care for the cattle, with every kind of human intercourse rendered perilous by the possibility of infection, the medieval Englishman obstinately carried on in his wonted way. Business was very far from being as usual but landlord and peasant alike did their best to make it so.

The simple structure of the more or less self-contained medieval village was, of course, far easier to maintain under stress than the elaborate social infrastructure of contemporary civilisation. So

far as the typical peasant was concerned, England's was a subsistence economy and to have let it founder would have been to
cease to exist as a society, almost, indeed, to cease to exist at all.
But the Englishman did more than just keep alive. Though the
Black Death violently distorted the pattern of village life,
wherever it was possible to do so taxes were paid and manorial
services rendered; the quick not only buried their dead but dutifully paid the fines on inheritance which were owing to the landlord. Within a few months one cell alone of Bruton Priory received fifty head of oxen and cattle as heriots; one for each tenant
who died. Here and there the burden was too great; organised
society ceased to exist for a few weeks or months, perhaps even
for ever. But such cases were the exception. By and large, and to
a greater extent than seems to have been true in continental
Europe, the fabric of society survived.

Was this a condemnation of the Englishman's timid conservatism, which led him to cling to his familiar chains when, by a
bold blow, he might have freed himself for ever? Or a triumph
for his durability and determination, his sturdy refusal to let himself be blown off course by an ill and transitory wind? The interpretation is a matter of taste and no formula could fail to be a misleading over-simplification. But it can at least be said that the
Englishman's reaction, or lack of reaction, was a victory for the
system under which he lived. It can be argued that, in the long
term, the Black Death struck a fatal blow at the manorial system
and heralded the end of the Middle Ages. Be that—for the moment—as it may; in the short term the Black Death provided an
impressive tribute to the system's strength and to the readiness of
the Englishman to accept the security which it offered and the
limitations which it imposed.

Judging by the rapid progress of the plague along the coast of
North Devon and Somerset, the infection travelled by boat by
way of the Bristol Channel as well as by the slower inland routes.
Whether it arrived first by land or water at Bristol is uncertain;
the latter, probably, though any port which was the centre of such
a busy traffic would have been an early victim in either case.
Bristol, the principal port of entry for the West Country, with
something close to ten thousand inhabitants, was the first im-

portant English city to be affected. 'There died,' recorded Knighton, 'suddenly overwhelmed by death, almost the whole strength of the town, for few were sick more than three days, or two days, or even half a day.'[31]

'Almost the whole strength of the town,' need not be taken too seriously, but it does seem that the plague was particularly ferocious in the city and its environs. Statistics must, as usual, be extrapolated from scanty evidence. There were ten new institutions for eighteen benefices—a figure which suggests that mortality among clerics was above the average for that part of England. The *Little Red Book* of Bristol lists the names of the town-council, the 'Forty-Eight,' for 1349. Of the fifty-two members which the 'Forty-Eight' whimsically contained, the names of fifteen had been struck through to show that they were dead. If all these died of the plague the mortality rate would have been a little under thirty percent; an unusually high figure for what must have been the cream of the city dignitaries. Things were undoubtedly a great deal worse in the crowded and stinking warrens in which the poor were forced to live. Boucher, the city historian, estimates an over-all death rate in Bristol of between thirty-five and forty percent and there is no reason to believe this figure exaggerated.[32] 'The plague,' according to an old calendar, 'raged to such a degree that the living were scarce able to bury the dead ... At this period the grass grew several inches high in the High St and in Broad St; it raged at first chiefly in the centre of the city.'[33] Cardinal Gasquet mentions the difficulties of the parson of Holy Cross de la Temple who had such urgent need to enlarge his graveyard that he took over an extra half acre without waiting for a royal licence. It is comforting to know that the King's pardon was subsequently forthcoming.

Meanwhile, Exeter had also been afflicted. According to one local historian of the nineteenth century, 'this dreadful calamity continued until the year 1357, when it happily ceased.'[34] Happily, indeed; but in fact there is no evidence to suggest that the plague in Exeter lasted longer than the usual span or that there was a renewed outbreak within the next few years. Another Exeter historian more prosaically states that the Black Death 'arrested the building of the cathedral nave ... paralysed our woollen trade and all commercial enterprise and suspended agricultural pursuits.'[35] Certainly, for two or three months, transportation in the area must have been reduced if not largely suspended, but Exeter, like every other English town in the fourteenth century with the possible exception of London, could live comfortably off the farms in its immediate neighbourhood. Though food may some-times have been hard to buy for want of middlemen to carry, prepare and sell it, there is no reason to think that the threat of famine was added to the city's miseries.

Inexorably the plague moved on through the West. It seems to have taken three or four months to complete its march but, by the middle of 1349, there can hardly have been a village in Devon and Cornwall which had not received its visit. At the isolated village of Templeton on the moors to the west of Tiverton there was no churchyard to accommodate the dead so that they had to be taken by cart-loads during the night to the mother church at Witheridge.[36] The deanery of Kenn to the south and south-west of Exeter is believed to have been the worst affected in the whole of England; eighty-six incumbents perished from a deanery with only seventeen parish churches.[37]

One casualty, luckier than most in that it survived, though seeming near to death at the time, was the tin industry of the west country. By the time of the plague the 'free miners' of Devon and Cornwall were a prosperous and powerful group enjoying a striking degree of local autonomy. The annual output of tin was some seven hundred tons. The death of many miners and the virtual disappearance of the market proved disastrous. In the years immediately following the plague production dropped away to almost nothing. As late as 1355 no tin at all was being produced in Devon. But in the more important mines of Cornwall recovery was more rapid and by the end of the century output had reached a peak which had only once been exceeded before the Black Death.[38]

EIGHT

Progress across the South

'Of which 22nd yeare and the next of the king's raigne is little to bee written, nothinge being done abrode, in effect, through the great mortality of the plague that raged all over the land; which as the historiographers of that time deliver, consumed nine parts in ten of the men through England, scarce leaving a tenth man alive.'[1]

For the historiographer concerned with the Black Death in England, 1349 is a year of which there is much to be written, for it was in the course of this year that almost every town and village was afflicted. At first it is possible to visualise the plague conducted, as it were, like a military operation. The initial attack on the Dorset ports; the bold thrust across country to the north coast so as to cut off communications between the western counties and the mainland; seaborne landings at points along the coast to outflank the defence; the slow mopping up of what opposition was left in Devon and Cornwall; and the main thrust towards the Thames valley and London. But after March, 1349, the analogy with a controlled campaign can no longer be pursued. To change the metaphor, the dykes were down and the water was everywhere. The infection no longer advanced regularly from point to point but sprang up simultaneously in a hundred places; reaching its peak, for no reason that can be established, in Norfolk and Suffolk before Cambridgeshire; in Hampshire before Surrey; in Warwickshire before Worcestershire. By July it was spreading across the northern counties, by the end of the year nowhere had been spared. Through winter and summer, through flood and drought; against old and young, weak and strong; the disease went imperviously on its way. To track its course with any precision would be a hopeless task: the most that is possible is to register its impact in the various regions and to highlight its workings at a few points where fuller detail is available.

When the plague turned eastwards after reaching the Bristol Channel the first city to be threatened was Gloucester. The town-council, horrified at the tragedy which had overtaken their neighbours in Bristol, decided to seek refuge in isolation. An embargo was placed on all intercourse between the two cities and the gates closed to any refugees who might carry with them the seeds of the plague.[2] But even if it had been possible to keep out every infectious human, and there is no doubt that the plague had been present in Bristol in its most virulent and infectious form, the citizens of Gloucester could have done nothing to protect themselves from the plague-bearing rat, making its way along the ditches or travelling in the river boats that plied up and down the Severn.

Bishop Wulstan Bransford remained on his country estate, occupied in the endless search for new priests to take the places of the dead. Between March and September alone eighty vacancies had arisen, almost all caused, according to the County History, by the death of the previous incumbent.[3] The Bishop himself died on 6 August, 1349. At Ham, a manor belonging to the Berkeleys not far from Cheltenham, so much land either escheated to the lord because of the tenants' death without an heir or was abandoned to look after itself that the bailiff had to hire the equivalent of 1,144 days' work to get the harvest in. The figure is impressive but so also is the fact that the extra labour was forthcoming, though no doubt at a considerable price.[4]

In so far as any pattern can be detected in the advance of the plague from West to East across England it seems to have struck from Bristol into Oxfordshire, Berkshire and Buckinghamshire and from Southampton and the West across Hampshire, Wiltshire, and Surrey towards London. The first prong of the advance provided some of the most fearful devastation of the whole epidemic.

The worst months for Oxfordshire were March, April and May, though there must certainly have been many cases before and after this period of maximum destruction. The county was, at this time, an archdeaconry within the vast diocese of Bishop Gynewell of Lincoln. Professor Hamilton Thompson's enormously valuable analysis of the Registers of the diocese shows that, so far as mortality among the clergy is concerned, the south of the county escaped lightly. In the deanery of Henley only a quarter of the beneficed clergy died and in Aston a mere 19%. But in the city of Oxford itself, 43% did not survive, in the deanery of Woodstock 42%, in Bicester 40% and in Chipping Norton 29%. In the whole archdeaconry just over 34% of the beneficed clergy perished, well below the 40.7% which was the average for Gynewell's diocese.[5]

Applying the suggested margin of error, one arrives at the highly speculative conclusion that total mortality in the arch-deaconry should have been between 25% and 37%. The Cartulary of Eynsham Abbey provides evidence to show that, in that part of the county at least, the lower of these estimates was well below the mark. [6] The Abbey itself, between Witney and Oxford, seems to have suffered badly enough. Abbot Nicholas had been deprived of his office by the Bishop because of some now for-gotten offence. Bishop Gynewell nominated two administrators to look after the Abbey pending the nomination of a new Abbot but, on 13 May, two of the senior brethren arrived to report that the first of his nominees was dead and the life of the second despaired of. He named in their place the two monks who had brought the news and sent them on their way. His new appoint-ments met with no greater success; both monks were dead before they reached the Abbey. In despair he forgave and reinstated Abbot Nicholas.

But the manors of the Abbey suffered still worse. Common tradition in England ascribes to the Black Death the responsi-bility for the disappearance from the map of many villages, leaving the church, usually the only solidly constructed building, as a solitary monument to the past. Certainly the Black Death helped the process of depopulation and so weakened many communities that they were unable to survive the economic and social vicis-situdes of the next two centuries. But very few villages can be shown to have been finally and completely deserted as a direct result of the Black Death.[7] One of the exceptions was the Abbot of Eynsham's manor of Tilgarsley (or Tilgardesle) where the collectors of the lay subsidy reported in 1359 that it was not possible to gather the tax because nobody had lived in the village since 1350. There is no reason to think that Tilgarsley was either rich or thickly populated before the Black Death but the fact that the tax was fixed at 94s 10d suggests that the community was reasonably prosperous or, at least, far removed from starvation.[8]

Another Eynsham manor, that of Woodeaton, went near to sharing the fate of Tilgarsley. 'In the time of the mortality of men or pestilence which befell in the year of our Lord 1349,' reads the Cartulary, 'scarce two tenants remained in the manor and they would have departed had not Brother Nicholas of Upton, then Abbot, . . . made an agreement with them and the other tenants who came in afterwards.' The bargain which the Abbot struck, giving the tenants a somewhat higher rent but less arduous feudal services, is an interesting example of the methods to which landlords were to have recourse in the years following the Black Death. He was only partially successful. By 1366 there were twenty-seven tenants in the village but six cottages still stood vacant.

At Cuxham, some seven miles south of Thame, only two reeves had been appointed to administer the manor in the whole period between 1288 and 1349. The old reeve died in March, 1349. His replacement died in April. His successor, a bailiff, died in June. His further successor died in July, and the fifth in line died or, at least, departed from the scene a year later in July, 1350. By 1360, the lord had given up any attempt to farm the manorial demesne and was seeking to put all his land out to rent.[9]

When the plague reached the city of Oxford, records Wood:[10] 'Those that had places and houses in the country retired (though overtaken there also), and those that were left behind were almost totally swept away. The school doors were shut, colleges and halls relinquished and none scarce left to keep possession, or make up a competent number to bury the dead.' The problem of what happened to the University during the Black Death is particularly bedevilled by suspect statistics. Richard Fitzralph, who had been chancellor a little earlier, recorded that 'in his time' there had been thirty thousand students but that, by 1357, the total had shrunk to a mere six thousand. He blamed the fall, however, not so much on the plague as on the machinations of the friars who lured students away by ignoble means.[11] Thomas Gascoigne, writing in the middle of the fifteenth century,[12] confirmed Fitzralph's figure, saying that he had seen the figure of thirty thousand cited in the rolls of the early chancellors as the student strength of the University. Wyclif raised the earlier total to sixty thousand and reduced the later to three thousand; not surprisingly attributing the mischief to the inflated worldly prosperity of the Church.[13]

Even in the bloated Oxford of the 1960's the total student body only numbers a little over ten thousand. No one to-day would accept as a reasonable estimate for 1348 a half or even a tenth of Fitzralph's figure, let alone of Wyclif's army of sixty thousand. Even at its peak of 1300 it is unlikely that the university held more than fifteen hundred students; three thousand would be the outside limit.[14] Given the number of potential chroniclers whom the University must have contained, it is curious how little evidence survives to show what happened to this population during the plague. If the experience of the larger monastic houses is any guide, then those students who elected to see out the epidemic from within their colleges paid heavily for their rashness. It is highly unlikely that they fared better than the townsman and probable that they did a great deal worse.

Berkshire was in a poor state even before the arrival of the plague. An exceptionally hard winter followed by sheep disease had set back the county's economy a few years before and, though things had improved by 1349, recovery was not complete. The impact of the plague was devastating but, except in certain areas, transitory. At Woolstone, almost on the borders of Wiltshire, the landlord in 1352 was forced to engage dairy women to do the milking and extra labour for weeding and most of the mowing. Yet by 1361 customary tenants were again established and paid labour largely dispensed with.[15] Of the Berkshire villages for which records exist, only in Windsor, a royal manor and as such likely to be given special treatment, were the changes introduced by the plague made permanent and all remaining villein services commuted for a money payment by 1369. Otherwise the pattern is one of losses made good, of a system strained but unbroken. Resilient and traditional, the manorial communities of England quickly put themselves back on an even keel and carried on, to the casual observer at least, as if the storm had never broken.

Buckinghamshire, where the Black Death was at its worst from May to September, does not produce a very different picture. In

Wycombe a startling 60% of the clergy died and it seems improbable that more than half the inhabitants stayed alive.[16] And
yet, by 1353, the town had recovered to the point that vacant
plots for building were being sought by would-be householders.
This however was true only of the town and not of the surrounding countryside. Wycombe's renewed prosperity did not filter
through to the Manor of Bassetbury on its outskirts where, even
fifty years later, the water-mill was in ruins, the fulling mill and
dye house untenanted, the barns of the manor in need of repair
and the tenants generally enjoyed larger holdings and payed
lower rents.[17] Meanwhile, at the manor of Sladen, near Berkham-
stead, in a deanery which suffered comparatively lightly, a jury in
August, 1349, declared that the miller was dead and his mill anyway valueless since there were no tenants left to need his services.
Rents to the value of £12 were no longer paid since all the
cottagers were dead. One cottage, where a certain John Robyns
survived and dutifully paid his 7s 0d a year, was the only part of
the manor deemed still to be of value.

One is left therefore with the curious situation that a town in
the centre of a deanery which lost almost as high a proportion of
its beneficed clergy as any in the country, had largely recovered
within three or four years, while a neighbouring manor was still
in difficulties fifty years later and another manor, in a part of the
county which seems to have been far less seriously afflicted, was
virtually wiped out. One moral to be drawn is that it is dangerous
to generalise even about relatively small areas—one village may
suffer disastrously; another, only a mile or two away, escape
virtually unscathed. Another moral, still more defeatist, is that all
statistics relating to the Middle Ages, particularly those deduced
by analogy or extrapolation, should be taken with a massive
pinch of salt.

But a partial and somewhat more rational explanation lies in
the nature of the different communities. A town like Wycombe,
if well run and energetic, could draw away labour from the surrounding countryside. Many of the surviving villeins in the
manors of the neighbourhood were disinclined to pick up the
shattered pieces of the rural economy. Others resented the efforts
of the landlords to exact feudal services which, in previous years
when labour was cheap and plentiful, had been allowed to lapse
or had willingly been excused against a modest money payment.

In a market town, anxious to encourage immigration so as to foster its thriving trades and commerce, such malcontents could find a welcome and, with luck, protection against any effort on the part of the former masters to restore the strayed sheep to its manorial fold. *Stadtluft macht Frei*, went the adage; and certainly many fourteenth-century villeins savoured their first breath of freedom in some country town seeking to restore the ravages of the Black Death. Wycombe regained its strength at the expense of neighbouring manors like Bassetbury; some at least of the lost tenants of Sladen were probably to be found at work in St Albans or Wendover. That the second half of the fourteenth century showed a progressive depopulation of the countryside is now almost a truism: that many towns showed a corresponding growth would be extremely hard to prove. But at least there can be little doubt that many of them, against the trend of population in the country as a whole, managed successfully to hold their own.

Meanwhile the southern prong of the plague's advance moved across Wiltshire and Hampshire. As in other areas, little points of certainty crop up above the mist of impressionistic vagueness. At Durrington, near Amesbury, eighteen out of forty-one tenants had disappeared by the end of 1349.[18] No rents of assize were paid at Tidworth. All the seven free tenants were dead on a moiety of the manor of East Dean and Grimstead and their lands were still standing vacant in 1350.

It would be possible to reel off a myriad such domestic details, each adding something to the overall picture but individually meaning little to the reader of to-day. To punctuate such a recital with constant ejaculations of dismay would be tedious to author and reader alike. But no study of the Black Death can make sense unless one constantly reminds oneself that this was not primarily a matter of statistics and social trends but of a shock of pain and appalling fear felt by many millions of people all over Europe. It is easy to say that medieval man lived closer to the threshold of death than his modern counterpart and that the impact of such wholesale destruction was therefore not so severe as it would have been to-day. But nothing could have prepared him for the horrors of 1348 and 1349. Behind the catalogue of bare ciphers, behind the laconic phrase '. . . because all the tenants were dead,'

lurk innumerable personal tragedies, little if at all less painful
because they seemed at that time to be the lot of all mankind.

The plague got a firm grip in Wiltshire before a significant
number of cases occurred in neighbouring Hampshire. The lists
of institutions to the benefices of Hampshire, assuming the usual
gap of a month to six weeks between mortality and replacement,
suggest that the first deaths took place at the very end of 1348,
that the worst months were February and March, 1349, and
that things were more or less back to normal by the end of the
year.[19]

But three months before the plague had struck, on 24 October,
1348, William Edendon, *alias* Edyndon or Edyngton, Bishop of
Winchester and former Royal Treasurer, had sent out warning
orders to all the clergy of his diocese.[20]

'A voice in Rama has been heard;' he lamented, 'much weeping
and crying has sounded throughout the various countries of the
globe. Nations, deprived of their children in the abyss of an unheard of plague, refuse to be consoled because, as is terrible to
hear, cities, towns, castles and villages, adorned with noble and
handsome buildings and wont up to the present to rejoice in an
illustrious people, in their wisdom and counsel, in their strength
and in the beauty of their matrons and virgins; wherein too every
joy abounded and whither multitudes of people flocked from afar
for relief: all these have already been stripped of their population
by the calamity of the said pestilence, more cruel than any two-
edged sword. And into these said places now none dare enter but
fly far from them as from the dens of wild beasts. Every joy has
ceased in them; pleasant sounds are hushed and every note of
gladness is banished. They have become abodes of horror and a
very wilderness; fruitful country places without the tillers, thus
carried off, are deserts and abandoned to barrenness.'

Whether the inhabitants of these erstwhile earthly paradises
would have recognised them from Edendon's description may be
doubtful but the picture of the fate which had overtaken them
must have caused dismay in the minds of all his readers. For, went
on the Bishop:

'. . . this cruel plague, as we have heard, has already begun
singularly to afflict the various coasts of the realm of England.

We are struck by the greatest fear lest, which God forbid, the fell disease ravage any part of our city and diocese. And although God, to prove our patience and justly to punish our sins, often afflicts us, it is not in man's power to judge the divine counsels. Still, it is much to be feared that man's sensuality which, propagated by the tendency of the old sin of Adam, from youth inclines to all evil, has now fallen into deeper malice and justly provoked the Divine wrath by a multitude of sins to this chastisement.'

To avert this doom the Bishop instructed his clergy to exhort their flocks to attend the sacrament of penance; on Sundays, Wednesdays and Fridays to join in saying the seven penitential and the fifteen gradual psalms and to take part, barefoot and with heads bowed, in processions around the market place or through the churchyards, reciting the greater litany. Three weeks later, while staying at Esher, he followed up this mandate with a further letter reminding the people 'that sickness and premature death often come from sin and that, by the healing of souls, this kind of sickness is known to cease.'[21]

But belated penitence availed nothing. The plague struck the diocese of Winchester with especial violence. 48.8 % of all beneficed clergy died, a figure not exceeded in any other diocese of England.[22] One explanation of this high mortality may be that the coast line of Hampshire was particularly exposed to ship-borne infection; the other two dioceses to suffer most were those of Exeter and Norwich, both of them similarly vulnerable. But it is difficult to make any sensible deductions valid for the whole of England about the factors which made any given area a ready target for the plague. In one region the hilly country seemed to suffer most, in another the plains. The fens of East Anglia escaped lightly, yet the valleys of the Severn and the Thames were devastated. The coast of Hampshire was much affected, yet Kent was relatively little damaged. Nowhere was immune but it seemed that only when the plague had come and gone could any town or county know whether or not it would prove especially susceptible.

In Crawley the population dropped from four hundred in 1307 to only a hundred and eighty in 1673. It did not reach four hundred again until 1851.[23] Certainly the Black Death was not alone responsible for what must have been a protracted process of depopulation. But the rapid changes in the methods used in the cultivation of the manorial demesne, in particular as regards the number of weekly workers, which immediately followed the epidemic show how much it must have affected the available labour force.[24] Prior to 1349 the reeve of Crawley, on behalf of his landlord, the Bishop of Winchester, was happy to receive 'fees for annual recognition,' that is to say, fees paid by villeins for the privilege of staying away from the manor to which they belonged. After this date no more such fees were received. Given the dearth of labour that then existed no landlord would willingly allow his villeins to deprive him of their services. Certainly the villeins did wander abroad, with greater frequency and success even than before the plague; but it was in defiance of their landlord and the law of the land.

Hampshire's off-shore islands suffered no less than the mainland.[25] The Isle of Wight was so reduced in population that, in 1350, the King remitted the tax due from the royal tenants. Almost every benefice in the island became vacant during the plague. Hayling Island, off Portsmouth, suffered quite as badly. 'Moreover,' said a royal declaration of 1352, 'since the greater part of the said population died whilst the plague was raging, now, through the dearth of servants and labourers, the inhabitants are oppressed and daily are falling most miserably into greater poverty.'[26] For these unfortunates, too, a reduced rate of taxation was conceded.

Winchester, the ancient capital of England, was as severely affected as any large town in the country. As usual it is difficult to establish either how large the population was before the plague or what percentage perished. Professor Russell has calculated that, in 1148, the population was about seven thousand two hundred and that, by 1377, the year of the poll tax, it had dropped to a mere two thousand one hundred and sixty.[27] Almost certainly numbers would have grown between 1148 and 1300 and dropped only slightly, if at all, between 1300 and 1348. The population at the latter date could not have been less than eight thousand and was perhaps as much as nine thousand or ten thousand. If one guessed that the Black Death killed four thousand people in the city the estimate would probably be conservative.

By January, 1349, deaths were running at such a level that the existing burial grounds were overcrowded. The Church insisted that all burials must take place in consecrated ground; the populace, more concerned with hygiene than theology, insisted with equal vigour that the bodies of the plague victims must be taken outside the city walls and buried in a common pit. When a monk from St Swithun's, the priory of the Cathedral, was conducting a burial service in the central churchyard, an angry crowd broke in and attacked and wounded him. The Bishop, outraged at this aggression by 'low class strangers and degenerate sons of the church' against a man 'whom, by his habit and tonsure, they knew to be a monk,' ordered the excommunication of the guilty. At the same time he gave the indignant citizens most of what they wanted—ordering the rapid enlargement of the existing grave-yards and the opening of new ones away from the centre of the town. He explained, for the benefit of the less well-informed members of his flock, that, since the Catholic Church believed in the resurrection of the dead, it was important that their corpses should be buried 'not in profane places, but in specially enclosed and consecrated cemeteries, or churches where with due reverence they are kept, like the relics of the Saints, till the day of Resurrection.'[28]

In the Middle Ages it rarely paid to get into a wrangle with a monk. The Bishop of Winchester had the last laugh when the time came to enlarge the churchyard of the Cathedral. With polite expressions of regret it was explained that this could only be done by reclaiming a stretch of land between the Cathedral and the High Street which had been granted to the priory by Henry I but subsequently 'usurped' by the Mayor, bailiffs and citizens as a site for a market and for bi-annual fairs. In Winchester, it was clear, either the quick or the dead were going to suffer and, if the Church had anything to do with it, it was not going to be the dead.

As in Siena, the plague left Winchester a tangible record of its visit. Edendon had formed grandiose plans for remodelling the west end of the Cathedral and reconstructing the nave in the Decorated style. He completed the demolition work in 1348, pulling down the two massive towers that flanked the Norman front. But when it came to rebuilding, the Black Death removed his labour force and funds ran short. A new west front was hurriedly flung up as a temporary measure until there was time and money to build something which would redound with greater éclat to the glory of Bishop Edendon. So far this makeshift has lasted something over six hundred years and still appears to have plenty of life left in it.

The plague reached Surrey, the other half of Bishop Edendon's diocese, a few weeks after Hampshire. March and April seem to have been the worst months. Banstead, four miles east of Epsom, was typical of many of the victims.[29]

The manor had been granted by Edward III to Queen Philippa as part of her dowry. A certain John Wortyng was installed as bailiff but evidently failed to win the confidence of the Queen's man-of-business. Some years after the Black Death had passed through the manor he claimed an allowance of £6 9s 10d for rent not paid on vacant tenements. The entry was struck out in his accounts and the sceptical note appended: 'Cancelled until inquiry is made into how many and what tenements are in the Queen's hands and for how much he could have answered on the issues of each tenement.' In the event he seems to have been proved justified. A jury sitting in 1354 found that twenty-seven out of a hundred and five villein holdings had been vacant since the epidemic. It is not unreasonable to deduce that a few others at least had found new tenants during this period and that the original death roll must therefore have included at least a third of Banstead's villeins.

The Black Death at Farnham has been the subject of a special study.[30] The hundred of Farnham was one of the richest and most populous in the great estates of the Bishop of Winchester. Judging by the Reeve's records in the Pipe Rolls there was a freakish first visitation of the plague at the end of 1348 which disappeared as mysteriously as it had come early in 1349 and was followed by the main outbreak at the same time as the rest of Surrey a few months later. In the twelve months between September, 1348 and September, 1349, one hundred and eighty-five heads of households died. The ratio between householders and dependants is a subject of some controversy[31] but, for the moment, it will be sufficient to assume that it could be no less than one to three. The total population of the hundred was between three thousand and four thousand; taking a figure half-way between the two, it would seem that some 20% of the inhabitants died.

The paradoxical result of this mortality was that the Bishop of

Winchester did very well financially. In a normal year fines paid on the estates of the deceased yielded between £8 and £20; in the twelve months of the Black Death this soared to £101 14s 4d. As heriots, the head of cattle which the heirs of every dead tenant had to hand over to the landlord, the reeve received twenty-six horses and a foal, fifty-seven oxen, one bull, fifty-four cows, twenty-six bullocks, nine wethers and twenty-six sheep. This windfall had its embarrassing side. Prices had slumped as a result of the plague and the reeve, even after killing and salting some of the oxen and cows, was forced to convert part of the demesne to pasture for the new herds.

On the debit side there was a substantial drop in rents; either because the tenants were dead or because conditions were so diffi-cult that all or part of the rent was remitted by the landlord. But, as on most of the manors of the Bishop, labour services were more important than a money rent. So great was the surplus of labour in Farnham immediately before the Black Death that it proved relatively easy to fill the vacancies and get in the harvest without much recourse to specially hired workmen. The three traditional harvest dinners were given for the twenty-four customary workers at a total cost of nine shillings, a figure very similar to that for earlier seasons. In the year in which the Black Death was at its worst, total receipts at Farnham were £305; total expenses only £43 5s 1¼d.

If this had been the whole story, then Farnham would have had cause to congratulate itself. But though the plague diminished in virulence it was still active. Between September, 1349, and September, 1350, another hundred and one head tenants died. By now the dwindling of the population must have meant that the ratio between householder and dependant also diminished but at least another three hundred villagers died. By the end of 1350, especially as a few further cases of plague occurred even in the last months of that year, more than a third of the people of Farn-ham must have been dead. Forty times in that year it was said that no fine was paid because there was nobody left to inherit. This meant that the cottage and land escheated to the lord; a situation which was profitable enough for the landlord in normal times when there were plenty of spare villeins to take up the tenement but disastrous when all the putative tenants were in their graves. The income from fines fell to £36 15s 10d and only four heriots were delivered, presumably remitted through charity or because the landlord had too many cattle already. By the end of 1349, fifty-two holdings were lying derelict. Thirty-six of these were filled up rapidly but the remainder proved more difficult. An increased amount of work on the demesne, particularly at harvest time, had to be done by hired labour and wages rose sharply in 1349 and 1350. With the virtual closure of the potters' and brick makers' industry in the neighbourhood, sales of clay and fern fell away to nothing. But even in this year the reeve could still show a reasonable profit on his operations.

It took some years to get things back to normal. Considerable pressure had to be brought on tenants to take up the vacant holdings but in the end all of them were filled. Wages never returned to the 1348 figure, but they soon fell below the inflated level of 1350. A market for clay and fern gradually reopened. Good administration; the support of a rich and powerful landlord and the natural wealth of the land, ensured that the hundred of Farnham, like the greater part of the Bishop's estates, was never a liability. In spite of the death of every third inhabitant, life and business went on much as before. In this Farnham was no more typical of England as a whole than the many manors already mentioned where the economy collapsed and income fell away to almost nothing. But its resilience was far from being unique or even exceptional. It is important to remember that both kinds of manors existed when seeking to establish a picture of England under the Black Death.

NINE

London: Hygiene and the Medieval City

And so the Black Death lapped at the gates of London. Compared with Paris, Vienna, Bruges or Constantinople, London may not have seemed so enormous a metropolis; certainly in architecture, painting and general grace of living Venice and Florence were far ahead. But it was still by a long way the most important commercial and industrial centre of England; three times, at least, as large as its nearest rival. Westminster, just outside the city walls, was the seat of government and of the King.

London seems to have grown more rapidly and more consistently than any of its rivals. Though the city was not included in the Domesday Book, at that time it probably had some fifteen or sixteen thousand inhabitants. By early in the thirteenth century, Professor Russell calculated, this figure must have doubled and, by 1348, doubled again to a population of some sixty thousand within the city wall.[1] The immediately outlying villages, integrated with the city in many ways and certainly part of the same unit from the point of view of the spread of the plague, must have added another ten or fifteen thousand to the total.

It would be inappropriate, in a book of this scope, to attempt any profound or detailed analysis of day-to-day life in a medieval city. Nevertheless there is much about the state of London, as for that matter about Paris or Florence, which is directly relevant to any study of the plague, since there were certain built-in features in the Londoner's pattern of life which contributed directly to its successful spread. Perhaps the most relevant of these was the overcrowding. Privacy was not a concept close to the heart of medieval man and even in the grandest castle life was conducted in a perpetual crowd. Hoccleve writes of an earl and countess, their daughter and their daughter's governess who all slept in the same room. It would not be in the least surprising to know that they slept in the same bed as well if, indeed, there was a bed. In the houses of the poor, where beds were an unheard of luxury, it would not have been exceptional to find a dozen people sleeping on the floor of the same room. In the country villages, indeed in many urban houses as well, pigs and chickens and perhaps even ponies, cows and sheep, would share the common residence. Even if people had realised that such a step was desirable it would have been physically impossible to isolate the sick. The surprise is not how many households were totally wiped out but, rather, in how many cases some at least of the inhabitants survived.

The dirt and inadequate sanitation of these hovels was, strictly speaking, less relevant to the spread of the Black Death. No one was going to become infected with bubonic plague by drinking tainted water or breathing foetid air. But, equally, it is true that the plague found its work easier in bodies weakened by dysentery, diarrhoea or the thousand natural shocks that the unclean body is particularly heir to. Still more important; warmth and dirt provide the ideal environment for the rat. The eventual victory of the brown rat over the plague-bearing black rat was in part due to the physical superiority of the former, but, at least as important, was a tribute to the rise in the standard of living and the substitution of brick for clay and wood which deprived the black rat of his sustenance and favourite way of life. The medieval house might have been built to specifications approved by a rodent council as eminently suitable for the rat's enjoyment of a healthy and care-free life.

What one might call the cinematic image of a medieval town is well known. Lanes barely wide enough to allow two ponies to pass meander between the steep walls of houses which grow together at the top, so as almost to blot out the light of day. The lanes themselves—they seem indeed more drains than lanes—are deep in mud and filth; no doubt to be attributed to the myriad buxom servant-wenches who appear at the upper windows and empty chamber pots filled with excrement on the passers-by. No street corner is without the body of a dead donkey and a beggar exhibiting his gruesome sores and deformities to the charitable citizens. Clearly one is in a society where hygiene counts for nothing and no town-council would waste its time supervising the cleaning of streets or the emptying of cesspools.

The picture, though of course over-drawn, is not entirely false. A medieval city, by modern standards, would seem a pretty filthy and smelly spot. But it would be unfair to suggest that citizens and rulers were indifferent to the nuisance or did nothing to remedy it. Thanks to the researches of Mr E. L. Sabine[2] and others, we now know much about conditions in London and the activities of the mayor, aldermen and common council. Though London, as the largest city of England, had the most serious problems, so also it had the greatest resources with which to deal with them. The overall picture of London's filth or cleanliness will be more or less valid for most of England's towns and cities.

Sanitary equipment, it need hardly be said, was scarce and primitive. In monasteries or castles, 'garderobes' were relatively common. Since 1307, the Palace of Westminster boasted a pipe between the King's lavatory and the main sewer which had been installed to carry away the filth from the royal kitchen. But this was probably unique in London; usually the privies of the aristocrats jutted out over the Thames so that their excrement would fall directly in the river or splash down the face of the castle wall. The situation was worse when the privies projected, not over a free flowing river but above a shallow stream or ditch. An inquest into the state of the Fleet Prison Ditch in 1355 revealed that, though it should have been ten feet wide and deep enough to float a boat laden with a tun of wine, it was choked by the filth from eleven latrines and three sewers. So deep was the resultant sludge that no water from Fleet Stream was flowing around the prison moat.

Occasionally citizens tried to dispose of their filth by piping it into the common drain in the centre of the street. A more ingenious technique was exposed at an Assize of Nuisances in 1347, when it was found that two men had been piping their ordure into the cellar of a neighbour. This ploy was not detected until the neighbour's cellar began to overflow.

Normally those fortunate enough to possess a private latrine would also have their own cesspool. In theory these had to be built to certain minimum standards; placed at least two and a half feet from a neighbour's land if they were stone-lined and three and a half feet if they were not. But there were many cases of seepage into adjoining properties and the contamination of private or public wells. Nor were these the only perils inherent in a cesspool, as the unfortunate Richard the Raker discovered when he vanished through the rotten planks of his latrine and drowned monstrously in his own excrement. Most blocks of tenement houses had their own privies though this was not invariable. But even where such facilities were lacking the chances were that there would be a public latrine not too far away.

Though sewers and cesspools were perhaps the most important of the common council's responsibilities, they provided by no means the only field in which the authorities saw reason to intervene. The three city butcheries of St Nicholas Shambles near Friars Minors in Newgate, the Stocks Market near Walbrook and East Cheap were subject to strict regulations. The years just before the Black Death, when cattle murrain was rife in the South of England, gave rise to many such prosecutions for selling meat described as 'putrid, rotten, stinking and abominable to the human race.' Offenders ran the risk of being placed in a pillory and having the putrid meat burnt underneath them.

The disposal of offal and other refuse was a serious problem. At the time of the Black Death the butchers of St Nicholas Shambles, had been assigned a spot at Seacoal Lane near the Fleet prison where they could clean carcases and dispose of the entrails. But, under pressure from the Prior of St John of Jerusalem, the site was moved and subsequently moved again, to a choice of Stratford or Knightsbridge; both suitably remote spots outside the city wall. 'Because,' as the royal instruction read, 'by the killing of great beasts, from whose putrid blood running down the streets and the bowels cast into the Thames, the air in the city is very much corrupted and infected, whence abominable and most filthy stinks proceed, sicknesses and many other evils have happened to such as have abode in the said city, or have resorted to it; and great dangers are feared to fall out for the time to come unless remedy be presently made against it . . .'[3] The final solution was to build a house on a pier above the Thames and dump the offal directly in the river during the ebb tide.

Even with such precautions the state of the streets was far from satisfactory. The tenement buildings, in which each story projected two or three feet beyond the one below, seemed designed for the emptying of slops, garbage and soiled rushes into the street. The gutters, which ran down the centre of the narrower streets and both sides of the wider ones, were generally inadequate to carry away the litter, augmented as it was by the dung of the innumerable domestic animals which lived in the centre of the city. The open sewers which ran down to the river were better able to manage the load but even these were often blocked and inadequate, especially in times of drought, to clear away all that was put in them.

To deal with these problems the common council appointed a number of 'scavengers' with instructions to 'remove all filth, and to take distresses, or else fourpence, from those who placed them there, the same being removed at their cost.' By 1345 the penalty for defiling a street had risen to two shillings and every householder was deemed responsible for a mess outside his house unless he could prove his innocence. At least one city raker was appointed for each ward and there seem to have been between forty and fifty carts and horses. The householders, knowing that they would be the ones to suffer if a street was allowed to grow filthy, could generally be relied on to support the efforts of the authorities. Sometimes, indeed, their aid seemed over-enthusiastic as when a pedlar threw some eel skins to the ground in St Mary-le-Bow and was killed in the resultant struggle.

But though refuse might have been removed with some efficiency from the city centre, too often the system subsequently broke down. Large dumps were established on the banks of the Thames and the adjoining lanes. In 1344 the situation had become so bad, especially around Walbrook, Fleet Stream and the city ditch, that a comprehensive survey of all the lanes was ordered. But though there was some improvement it does not seem to have lasted long. Thirteen years later the King was complaining bitterly that his progresses along the Thames were being disturbed by the 'dung, lay-stalls and other filth' which were piled up along the bank.[4]

The overall picture, therefore, is of a city squalid and insanitary enough but aware of its deficiencies and doing its best, though with altogether inadequate tools, to put things right. The records reveal many cases of behaviour in wanton defiance of the rules of hygiene but the very fact that such behaviour was commented on and sometimes prosecuted shows that the picturesque excesses, so dear to the heart of the antiquarian, were not permitted to flourish unchecked. A responsible city council and a population on the whole aware of its civic duties did quite a good job of keeping London clean.

But the Black Death proved altogether too much for the public health services. In 1349, the King wrote to the mayor to remonstrate about filth being thrown from the houses so that 'the streets and lanes through which people had to pass were foul with human faeces and the air of the city poisoned to the great danger of men passing, especially in this time of infectious disease.' The mayor was helpless. Not only had many of the efficient cleaners died or deserted their post and the machinery for the enforcement of the law been strained beyond its capacities but also the technical problem of transporting something over twenty thousand corpses to the burial grounds had imposed an extra and unexpected burden on the skeleton force which remained. Even ten years later the service was far from normal; in the year of the Black Death itself, the most lurid imaginings of a romantic novelist would hardly have done justice to reality.

There are so many different routes by which the Black Death could have arrived in London that it would be pointless even to speculate from whence it came. By the end of September, 1348, the Prior of Canterbury had addressed an alarmed mandate to the Bishop of London on the incursions which the plague was making in the latter's diocese,[5] but it does not follow from this that infection was already within the city. Nevertheless it seems certain that the city was affected before the greater part of the surrounding countryside and that there were cases as early as November, 1348, and perhaps sooner still. But the main force of the epidemic was not felt until the beginning of the following year.[6]

Spring, though any generalisation about the plague has many exceptions, was usually one of the less dangerous seasons for outbreaks of bubonic plague.[7] Both from the histories of subsequent epidemics in London and from the evidence which survives of the Black Death itself it seems likely that, from January to March, a strain of pulmonary plague predominated but that the pure bubonic plague came into its own with the warm weather in the late spring and summer. As always in a large community, the disease lasted longer and consumed more gradually than in a small town or village. Deaths were still common till far on into 1350 and, though the full fury of the epidemic lasted only three or four months, almost two years passed between the Black Death's arrival and the final casualties.

In January, 1349, shortly before Parliament was due to assemble, the King prorogued it on the grounds that '. . . the plague of deadly pestilence had suddenly broken out in the said place and the neighbourhood, and daily increased in severity so that grave fears were entertained for the safety of those coming there at that time.' The King's concern for his legislators was proper and, in the event, well justified but seems a little premature. It may well have been little more than a pretext.[8] In January, 1348, Parliament had proved recalcitrant and, when they eventually and grudgingly granted a subsidy for three years, they made it clear that they felt the burden of taxation to be unreasonably heavy. With his subsidy safely in his pocket the King would have jumped at an excuse to be spared the grumbling of his legislators. The epidemic came just in time to furnish it.

The existing graveyards were soon too small to meet the demand. A new cemetery was opened at Smithfield and hurriedly consecrated by Ralph Stratford, Bishop of London. But the second of the two new cemeteries, founded by the distinguished soldier and courtier, Sir Walter Manny, has provided historians with the greatest confusion. Early in 1349, Sir Walter leased for twelve marks a year and subsequently bought some thirteen acres of unused land to the north-west of the city walls at a spot called Spittle Croft. He built a chapel on the site, dedicated to the Annunciation, and threw it open for the overflow of victims of plague within the city.[9] Eventually the Charterhouse was built on part of the ground. The confusion arises over the number of dead which the new graveyard accommodated. Robert of Avesbury says that two hundred people were buried there almost every day between the feast of the Purification (2 February), and Easter (2 April).[10] If this is taken to mean that burials took place at this rate during what must have been the worst months of the plague and it is assumed that they continued at a reduced rate for the next few months, then a minimum of seventeen or eighteen thousand victims must have found a home there. This figure is enormous but still trifling compared with that put forward by the London historian, Stow,[11] who recorded that he saw an inscription in the churchyard which read:

'A great plague raging in the year of our Lord 1349, this churchyard was consecrated; wherein, and within the bounds of the present monastery, were buried more than fifty thousand bodies of the dead, besides many others from thence to the present time, whose souls God have mercy upon. Amen.'

He claimed that this figure had been confirmed by his study of the Charters of King Edward III.

Camden claimed to have seen the same inscription but, in his recollection, the figure was an almost equally startling forty thousand. There is no indication that these new cemeteries were intended to replace rather than supplement the existing churchyards, one and probably two other new ones were opened; it would seem therefore that Manny's site could not possibly have taken more than half the victims of the plague, and probably a great deal less. If Stow's figures were correct, this would mean, therefore, that a minimum of one hundred thousand people died of the Black Death in London, a figure credulously adopted by Rickman in his Abstract of Population Returns.[12] Even if Robert of Avesbury's figure were accepted the total of the dead could hardly be less than forty thousand. Figures above fifty thousand have frequently been bandied about. Yet all these totals seem unreasonably high when set alongside a population of sixty or seventy thousand. The ecclesiastical registers, which might have provided a more accurate check, do not survive but such snippets of information as exist—as, for instance, that three out of seven benefices in the gift of the Abbey of Westminster fell vacant in the spring and summer of 1349 and both those in the gift of the Abbey of St Albans—suggest that casualties in London were more or less in line with those in other cities. Certainly there is no reason to think that it suffered less. A total of between twenty and thirty thousand dead, probably closer to the higher figure, is likely to be as accurate a guess as one will get unless some further source of statistics is discovered.

Though, as everywhere, the poor suffered most, there were quite enough deaths among the rich and powerful to show that nobody was immune. John Stratford, Archbishop of Canterbury, died at his manor of Mayfield in August, 1348. It is quite probable that he was not a victim of the plague but there is no doubt about his successor, the Chancellor, John Offord, who died in May, 1349, at Westminster before he could even be enthroned. Clement VI then appointed the great scholar Thomas Bradwardine but, he, in his turn, died in the Bishop of Rochester's London palace on 26 August, 1349. A former Chancellor, Robert Bourchier, died of the plague in London and one of his successors, Robert Sadington, died in 1350, though of uncertain cause. The royal family seems to have kept out of trouble, the only casualty being the King's daughter Joan who died at Bordeaux on her way to Portugal, but Roger de Heyton, the royal surgeon, died on 13 May, 1349.[13] There were heavy losses among the dignitaries of the City. All eight wardens of the Company of Cutters were dead before the end of 1349. Similarly, the six wardens of the Hatters' Company were all dead before 7 July, 1350, and four wardens of the Goldsmiths' Company died in 1349.[14]

The great Abbey of Westminster did not escape. Simon de Bircheston, its truculent Abbot, who had been prosecuted for assaulting a royal stonemason some twenty years before, took refuge in his country home at Hampstead. But in spite of his precaution he was an early victim[15] and twenty-seven monks accompanied him to the grave. A large black slab in the southern cloister of the Abbey probably commemorates their death and may even cover their remains.[16] By May, Simon Langham, who had been appointed a prior only a month before, was the only monk left deemed fit to administer the monastery.

The many deaths in the countryside and the natural reluctance of the carters to venture into the inferno of London meant that the usual supplies of food often failed to arrive. The Black Death prevented anything near a famine developing by rapidly and substantially reducing the demand but it was still often difficult for a citizen of London to know where to turn for his next loaf of bread. Piers Plowman noted:

> 'It is nought long y-passed
> There was a careful commune
> When no cart com to towne
> With bread fro' Stratforde.'

Many Londoners went out into the adjoining countryside in

search of food and so spread the plague among those who had sacrificed a profitable market in the hope of escaping it.

London survived. Probably, indeed, it recovered as quickly as any city in England. In 1377, the population of the city itself seems only to have been about thirty-five thousand but this was after further attacks of plague and takes no account of population growth in the immediate vicinity.[17] The fact that all the chancery and exchequer work continued to be done in London was a powerful magnet and there was no city in which the villein, anxious to escape the attentions of a vengeful lord, could bury himself with greater confidence. Dr Creighton probably goes too far when he says that we may be sure 'from all subsequent experience; that the gaps left by the plague were filled up by influx from the provinces and from abroad in the course of two or three years,'[18] but it is likely that, even in so short a period, much of the lost ground was made up.

And yet the mark left by the Black Death was not to be seen only in the new cemeteries. The sharp fall in moral standards which was noticed in so many parts of Europe in the years after the Black Death was nowhere more striking than in London. Such accusations of degeneracy recur throughout the ages—this time they may have had slightly greater justification than usual. Knighton reported that criminals flocked into the city[19] and John of Reading told of the great increase in crime, in particular crimes of sacrilege.[20] From this period, the city began to enjoy a doubtful reputation in the eyes of other Englishmen—a city of wealth but also wickedness; of opportunity, but opportunities to earn damnation as well as fortune. Walsingham denounced the Londoners roundly: 'They were of all people the most proud, arrogant and greedy, disbelieving in God, disbelieving in ancient custom.' Those who live in great cities are traditionally believed to be harder, more sophisticated and more rapacious than their country cousins but the Londoner certainly acquired his reputation the hard way and probably went a long way to deserving it. Any city which suffers as London suffered and rebounds rapidly to even greater prosperity can be excused a certain fall from grace during the years of its recovery.

TEN

Sussex, Kent and
East Anglia

Sussex and Kent were assailed from every side as the Black Death moved across from the West, spread south from London and made its independent entry at half a dozen Channel ports. In Sussex the absence of episcopal registers makes any generalisation difficult. A few details, selected more or less at random, show that the county did not escape more lightly than its neighbours. Only five brethren survived out of a total strength of thirteen at the Priory of Michelham.[1] The Abbot of Battle had secured permission to fortify his Abbey only ten years before, but no moat or ramparts could save him from the plague which carried him off with more than half his monks.[2] At Appledram, almost in Hampshire, the number of customary reapers was reduced from two hundred and thirty-four to one hundred and sixty-eight, a drop of 28%; at Wartling, right at the other end of the county, twelve freemen and villeins died in March, 1349, and a further sixty had perished by October, twenty-five leaving no direct heir.[3]

For the Black Death in Kent there has fortunately survived the account of William of Dene, a monk of Rochester, one of the few English chroniclers who handles the events of the day with anything like the impressionistic brio of his continental counterparts.[4]

'In this year,' he recorded, 'a plague of a kind which had never been met with before ravaged our land of England. The Bishop of Rochester, who maintained only a small household, lost four priests, five esquires, ten attendants, seven young clerics and six pages, so that nobody was left to serve him in any capacity. At Malling he consecrated two abbesses but both died almost immediately, leaving only four established nuns and four novices. One of these the Bishop put in the charge of the lay members and the other of the religious, for it proved impossible to find anyone suitable to act as abbess.

'To our great grief,' went on the monk, 'the plague carried off so vast a multitude of people of both sexes that nobody could be found who would bear the corpses to the grave. Men and women carried their own children on their shoulders to the church and threw them into a common pit. From these pits such an appalling stench was given off that scarcely anyone dared even to walk beside the cemeteries.

'There was so marked a deficiency of labourers and workmen of every kind at this period that more than a third of the land in the whole realm was let lie idle. All the labourers, skilled or un-skilled, were so carried away by the spirit of revolt that neither King, nor law, nor justice, could restrain them . . .

'During the whole of that winter and the following spring, the Bishop of Rochester, aged and infirm, remained at Trottiscliffe [his country manor between Sevenoaks and Rochester], bewailing the terrible changes which had overcome the world. In every manor of his diocese buildings were falling into decay and there was hardly one manor which returned as much as £100. In the monastery of Rochester supplies ran short and the brethren had great difficulty in getting enough to eat; to such a point that the monks were obliged either to grind their own bread or to go without. The prior, however, ate everything of the best.'

The chronicler probably had his facts right so far as the details of the Bishop's establishment were concerned but his qualifica-tions are less impressive as an observer of the agricultural scene in Kent or, still more, 'the whole realm.' The reference to 'more than a third of the land' lying idle had as little statistical signific-ance as similar statements that half or three-quarters of the popu-lation were dead. It was certainly exaggerated; but quite as certainly it was a tribute to the very real dislocation which had afflicted the farming lands of Kent and the forlorn and unkempt air which they must often have presented.

The priory and convent of Christ Church, Canterbury, suffered little in the epidemic, only four inmates dying. Thorold Rogers attributes this to the good water supply and efficient drainage system which had been installed by an earlier prior. It is hard to accept any very direct relationship between bubonic plague and pure drinking water,[5] but the fact that the establishment was clean and free from rats would certainly have helped to keep infection at bay. It is surprising that the flow of pilgrims to Canterbury hardly slackened even when the plague was at its worst.[6] It might have been supposed that the mortality among putative pilgrims and the obvious perils of travel in time of pestilence would have been enough to scare off even the most devout. Presumably those who had already survived an outbreak wished to give thanks for their deliverance while those who were yet to experience one hoped to accumulate merit in advance. In neither case was the result likely to be wholly satisfactory. No doubt their determination was welcome in Canterbury, since every visitor was a source of income, but as each wave of arrivals brought in fresh infection even the most avaricious of citizens must have asked himself whether the blessing was an unmixed one.

Somehow the old and decrepit Bishop Haymo of Rochester managed to survive while all his retinue perished around him. Reference has already been made to the long-established belief that the Black Death struck down the strong in the prime of their life and spared the children and the aged.[7] In this, as in all things, it would have been like the Bishop to seek to conform to tradition. But such evidence as exists does not support the theory. Analysing five hundred and five inquisitions post-mortem, Professor Russell found that by far the heaviest mortality, 46%, occurred in one of the most senior age groups, those between fifty-six and sixty.[8] The oldest age group of all, those above sixty, suffered next most badly, with mortality of almost 40%. It is, of course, true that a higher proportion of this group would anyhow be likely to die within any given period even without the Black Death to help them on their way. But the incidence of mortality is still strikingly higher than among those in the prime of their life: only 20% for those between twenty-one and twenty-five; 19% for the twenty-six to thirties; and 28% for the thirty to thirty-fives. Tradition does seem to be right, however, in maintaining that the children were spared. Only 7% died of those between six and ten and 15% between eleven and fifteen.

By these rules the Bishop was fortunate to have outlived his priests and outstandingly lucky to have survived his pages. After such hazards one might think that he had earned a tranquil old age. But new troubles beset him. Even when the worst of the plague was over he found that his clergy hesitated to do their duty in their parishes; preferring either to remain prudently at home or to abandon their flock to their fate and retreat to what they hoped might prove a safer area. The phenomenon was by no means peculiar to Kent but it seems to have been particularly remarked on there. Stephen Birchington of Canterbury also referred to the fact that '. . . parish churches remained altogether unserved, and beneficed parsons turned away from the care of their benefices for fear of death . . .'[9]

Dene of Rochester in his turn commented on the decadence which had overtaken the country under the stress of the Black Death. His strictures on workmen possessed by the spirit of revolt have already been noted but worse was to follow: 'The entire population, or the greater part of it, has become even more depraved, more prone to every kind of vice, more ready to indulge in evil and sinfulness, without a thought of death, or of the plague which is just over, or even of their own salvation . . . So, day by day, the peril in which the souls of clergy as well as people are to be found has grown more dangerous . . .'

One sees Dene of Rochester as a crabbed, reactionary figure, passed over for promotion, embittered, filled with the darkest doubts about the younger generation. But there are too many similar reports to leave room for doubt that he had grounds for his grumblings.

There is not much to delay one in the other counties of the south. In Hertfordshire, March and April were the worst months. But many cases occurred even when the summer was over and there seems to have been a second, milder outbreak in the course of 1350. In certain manors, where the ravages of the plague had been particularly ferocious, it became the custom to head future schedules of expenditure with 'an enumeration of the lives which were lost and the tenancies which were vacated after the great death of 1348.'[10] The archdeaconry, an area considerably larger than that of the county, does not seem to have suffered particularly badly. In the low-lying fen district, St Ives lost only 23%, Holland 24% and Peterborough 27% of the beneficed clergy.[11] Marshes and fens, perhaps because of their association with mosquitoes, are generally linked in the popular mind with fever and disease. In the case of the Black Death they belied their reputation. One possible explanation is that they were often remote from the sea and the main lines of communications and therefore to some extent sheltered from infection. Another is that such damp and sparsely inhabited areas held little appeal for the wandering rat. But the impact of the plague was too spasmodic to allow even exceptions to the general rule to be defined exactly and certain areas of fen country suffered as badly as any in the country.

To quote statistics which show that Hertfordshire suffered less severely than other counties is not to detract from the agonies which the inhabitants endured: to the victims it mattered remarkably little whether the mortality was 37% or a mere 34%, the risk and the pain of death seemed much the same. A scrawl on the wall of the church of St Mary, Ashwell, somehow catches the black horror of the plague. 'Wretched, terrible, destructive year . . .' the unknown scratched in the stone sometime in 1350, '. . . the remnants of the people alone remain . . .'[12] There are plenty of examples in the county of almost complete disaster. At Standon, six miles north of Ware, thirty-two customary tenants were supposed to mow the lord's hay. In 1349 no men went to mow, and the hay was left to rot in the fields.

But where an analysis has been made of a group of manors big enough to provide a reasonable sample, one is once again struck with the amazing speed of the countryside's recovery. Dr Levett, whose individual contribution to the history of the Black Death, though in some respects now seen to be too extreme in its conclusions, has done so much to bring sanity and cool scientific reasoning into a sphere peculiarly rich in ill-supported fantasy, has studied, *inter alia*, the manors of the great abbey of St Albans.[13] The plague, it is true, was at its worst on these manors in April and May, which was the least dangerous period from the point of view of the rural economy. The corn crops could safely be left to take care of themselves until what was left of the population had time to attend to them, and weeding and ploughing dispensed with for one year without serious consequences. But, granted this minor stroke of luck, a serious setback to the manors' economy might have been expected. So indeed there was, but as Dr Levett puts it:

'The average historian of the plague period seems to have worked from two assumptions:

'(1) that every peasant farmer was occupied to the utmost of his capacity before the pestilence; and (2) that after it the whole remaining population, supine and unalert on their own holdings, tended to rise up and wander about the country in search of high wages. Neither assumption will hold water.'

Once the worst of the shock was over, energy, discipline and intelligent administration quickly got the wheels of agriculture turning once again. The St Albans' manors were well-run and prosperous, on good farming land and with the power and wealth of the Abbey to sustain them. They had little difficulty in luring away labour from other, less fortunate estates. It would be a great mistake to assume that what was true of them was true of the majority of English manors but, equally, they were by no means unique.

The Abbey itself suffered rather worse than its manors.[14] Michael of Mentmore, one of the greatest of its Abbots, was struck down after thirteen years in office on 2 April, 1349:

'. . . being the first to suffer from the dread disease which was later to carry off his monks. He began to feel the first symptoms on Maundy Thursday, but out of reverence for the festival and remembering our Lord's humility, he celebrated High Mass and then, before taking his dinner, humbly and devoutly washed the feet of the poor. After he had taken his dinner he proceeded to wash and kiss the feet of all the brethren and to carry out all the offices of the day alone and without assistance. The next day when his sickness became worse, he took to his bed and, as a true Catholic, made his confession with a contrite heart and received the sacrament of extreme unction. Amidst the sorrow of all who surrounded him he endured until noon on Easter Day . . . And there died at that time forty-seven monks . . .'[15]

The neighbouring archdeaconry of Bedford lost 38.6% of its beneficed clergy.[16] At Millbrook, to the south of the town of Bedford, the lord of the manor Peter de St Croix was among those who died. At the inquisition it was said that all the bondmen and cottars were dead and, a few months later, his son and heir Robert followed his father.[17] With incidents such as this a commonplace it is not surprising that Bedford, whose main function was to serve as a centre for the agricultural lands around it, lost drastically in prosperity. Certainly the general decline of the English rural economy began long before 1350 and would probably have continued even if there had been no epidemic but, in the case of Bedford at least, the County History believes the plague to have been the decisive factor. It took the town a hundred and fifty years to recover its former strength.

In East Anglia, arbitrarily defining this somewhat fluid concept to include Cambridgeshire as well as Norfolk, Suffolk and the north of Essex, the Black Death seems to have arrived in March, 1349, reached its peak in May, June and July and died out during the autumn.[18] A typical case was that of the manor of Cornard Parva where nine deaths were recorded at the Court held on 31 March. By 1 May another fifteen were dead of whom seven left no heir; leaving the presumption, though by no means the certainty, that in such cases an entire family had been exterminated. By 3 November, after a long gap during which, presumably because of the plague, no Court was held, the parson and a further thirty-six tenants were dead, this time twenty leaving no heir. The lasting economic damage done by the Black Death was demonstrated in the market town of Sudbury in Suffolk. In 1340 there were one hundred and seven ancient stalls licensed for the weekly market. By 1361 the number had dropped to sixty-two.[19]

Yarmouth at this period was one of the most flourishing towns and certainly the leading sea-port of East Anglia. Seebohm believed that the population in 1348 must have been more than ten thousand[20] and Cardinal Gasquet, pointing out that Yarmouth had two hundred and twenty ships and furnished three times more sailors than London for the attack on Calais, argued that this estimate must be much too low.[21] Modern research, which almost always seems to lead to the reduction of earlier estimates, would on the contrary consider it decidedly too high. But the plague certainly wreaked terrible havoc and recovery was slow. At the beginning of the sixteenth century a petition of the burgesses to Henry VIII referred to the great pestilence: 'by reason whereof the most part of the dwelling places and inhabitants of the said town stood desolate and fell into utter ruin and decay, which at this day are gardens and void grounds . . .' The unfinished tower of St Nicholas, begun in the age of prosperity and abandoned when money and labour were alike lacking, pays tribute to the thoroughness with which the Black Death did its work.

The bailiwick of Clare, about fifteen manors belonging to the Earls of March and scattered widely over East Anglia, provides an interesting illustration of the extent to which a great landlord could hold his own in time of trouble.[22] The bailiwick on the whole was not very seriously affected: certainly much less so than another group of Mortimer manors around Bridgewater Castle in Somerset. Standon, a village not in East Anglia at all but in Hertfordshire, was the worst hit. The number of labour services was halved and some tenements remained unoccupied for more than twenty years. But even on this battered manor, once the initial shock had been weathered, the total of rents received fell only by some 15%. Wages rose sharply in the immediate aftermath of the Black Death but by the 1360s had been pegged back to little above the level ruling before the plague. For a year or two a large part of the lord's demesne was left uncultivated but this too was soon put right.

The Mortimers seem to have suffered decidedly less than most of their neighbours. This was, as a generalisation, true of the whole class of great territorial magnates who held their own throughout the recession of the fourteenth and fifteenth centuries. Dr Holmes has pointed out in his analysis of the estates of the higher nobility that, by thus retaining their habitual revenues, they were in effect taking a larger share of a decreasing product.[23] The damage thus done was blanketed in the middle of the fourteenth century by the reduced pressure on resources which the great mortality of the Black Death produced. But, in the long run, it was a movement against an economic trend which was to weaken the stability of the system which made it possible.

Norfolk in particular is rich in those holes and mounds and crumbled ruins which local historians have painstakingly identified as the lost villages of another era. We have already referred to the tendency to ascribe such disappearances to the baleful effects of the Black Death. In Norfolk this seems even less justified than elsewhere. More than a hundred Norfolk villages existed at the time of the Domesday Survey and have subsequently disappeared. Thirty-four of these had already vanished by 1316 and others almost certainly fell into desolation between that date and 1348. Only one, Ringstead Parva, ceased to exist as a community between then and 1351; the bulk of the remainder lingered on till the second half of the fifteenth century or even later.

'There can be little doubt,' concluded Mr Allison,[24] 'that the Black Death played no more than a contributory part in village depopulation in Norfolk.' But though it is easy to exaggerate the destructive power of the plague, it would be far more misleading to minimise it. Great tracts of the county, in particular the barren Brecklands around Brandon, can literally be said never to have recovered. Economically weak and on the decline even before 1348, the blow which the Black Death then inflicted and, more insidious, the opportunities which the plague created for finding better land in other, more prosperous parts of the country, sealed the doom of these unlucky areas. The hold which civilisation had established on the border-lands during the great expansion of the thirteenth century was especially weak in East Anglia. Now it was driven inexorably back and the wild took over its own again.

Bishop Bateman was conducting peace negotiations with the French when the plague neared the borders of his diocese. He returned by sea to Yarmouth on 10 June and was told on landing that his brother, Sir Bartholomew Bateman of Gillingham, was already in his grave.[25] Hurrying to Norwich, he found the plague raging; his Vicar General, Thomas de Methwold, lurking at Terlyng in Essex; and his palace, next to the new cemetery in the Cathedral Close, made almost intolerable by the stench of the dead. He at once ordered Methwold to return to his duties but barely was the Vicar General back at work than the 'intrepid Bishop,' as Dr Jessop hopefully called him, was himself on his way to his rural manor at Hoxne nearly twenty miles south of Norwich. He spent three days at Ipswich in the next few months but did not visit the capital of his diocese again till the plague was safely over.

Certainly Norwich was anything but a pleasant place to be in the summer and autumn of 1349. 'There died,' recorded Blomefield, possibly confusing figures for the city with those for the county or diocese, 'no less than 57,104 (or more rightly as others have it, 57,374) persons in this city only, besides religious and

beggars.'[26] He admitted that this figure might seem surprisingly
high since the population of Norwich when he wrote in 1806 was
still quite a lot less than the casualties of 1349, but explained that
the city, in the middle of the fourteenth century, was 'in the most
flourishing state she ever saw, and more populous than she hath
been ever since.' Given about a thousand inhabitants to each of
the sixty-odd parishes and throwing in an allowance for the sub-
urbs and the religious houses, Blomefield calculated that the total
population of 1348 must have been a minimum of seventy thousand. Seebohm accepts that the death rate may have been in the
neighbourhood of fifty-seven thousand but reduced the total
population to sixty thousand; evidently feeling that a death rate
of 95 % called for no special explanation.[27] Basing himself on the
Leet Rolls Professor Russell estimates that the population of Norwich was some thirteen thousand in 1311.[28] Subtracting the poll
tax figure for 1377 of five thousand nine hundred and twenty-
eight, the figure for the population which remains to be accounted for comes remarkably close to seven thousand one hundred and four; the last four figures of Blomefield's total. Russell
ingeniously surmises that the five at the beginning was added by
some careless transcriber and that the ancient record to which
Blomefield referred should in fact have been read as a broadly
accurate seven thousand dead. He could be right but since, as he
himself frequently points out, medieval statistics were invariably
grossly over-stated, it seems easier to believe that the usual rule
applied in the case of Norwich. What at all events seems certain is
that Norwich, the second city of the kingdom, lost more than
half its population and not only never recovered its position
in relation to the rest of England but, in absolute terms, had
barely regained its vanished citizens by the end of the sixteenth
century.

Though the whole of Bishop Bateman's diocese may not have
suffered as much as its capital there is no doubt that mortality
among the clergy was unusually high. In the years before the
Black Death the average annual figure for episcopal institutions
was eighty-one. In the year between 25 March, 1349 and 25
March, 1350, the total rose to eight hundred and thirty-one.[29]
For the population as a whole, Jessop remarked: 'If any one
should suggest that *many more* than half died, I should not be disposed to quarrel with him.'[30] It is unlikely that he would go so

far if he were writing to-day but at least it seems certain that the
death rate in East Anglia was well above the national average.

The Bishop of the neighbouring diocese of Ely was in Avignon
when the plague reached his territories. He seems to have made
no effort to return. He had already appointed five Vicars General
to look after the diocese in his absence and, on 9 April, appointed an additional three 'to exercise their duties until death.'[31]
In his letter he referred to 'the epidemic, as it is called, won-
drously increasing in the diocese,'[32] a phrase soon to be fully
justified but at the time a little premature since it was not till
April that the plague took a serious grip in Cambridgeshire. He
specified which of the Vicars General was to dispose of vacant
benefices in his absence and carefully listed the order of succession
to this attractive prerogative.

In the whole diocese there were eighteen times as many
institutions as in a normal year, a far higher ratio even than in
Norfolk and Suffolk. Only one of these institutions is known to
have been due to the resignation rather than the death of the
previous incumbent though there are other cases in which
details are lacking.[33] But the incidence of the plague seems to have
been even more erratic than in other parts of England. Within a
radius of ten miles of Cambridge thirty-five out of fifty tenants
died on the Crowland manors at Oakington, twenty out of forty-
two at Dry Drayton, thirty-three out of fifty-eight at Cottenham
and yet at the manors of Great Shelford and of Elsworth, though
there may have been deaths which were not recorded in the Court
Rolls, there is no evidence that the Black Death had any effects at
all. All these communities were in broadly similar country with
populations of between two and four hundred. Their inhabitants
suffered the same weather, farmed the same land, ate the same
food. No one, be he epidemiologist or historian, has yet been able
to suggest plausible reasons why one manor should have lost half
its tenants while another seems to have suffered little if any ill
effects.

There is remarkably little to show how severely the plague
affected the University at Cambridge. The only College which
furnishes any useful material is King's Hall where sixteen out of
forty resident scholars died between April and August. But one

can deduce that the losses among students must have been severe by what is known of events in the town of Cambridge. Though some of the damage may have been done by the second epidemic of 1361, an idea of the devastation is given by a letter from the Bishop of Ely written in 1366 which suggested the amalgamation of two of the city parishes on the grounds that there were not enough people left to fill even one of the churches. Practically the whole of the town on the Castle side of the river, 'The Ward beyond the Bridge,' seems to have been wiped out. Most people in All Saints in Castro also died and the few parishioners left alive moved out. The nave of All Saints' Church fell into ruins, leaving 'the bones of the dead exposed to the beasts.'[34]

If eight hundred beneficed clergy died in East Anglia, well over twice as many priests must in all have perished. Taking into account subsequent desertions there must have been at least two thousand vacancies; nearly two thousand five hundred if Cambridgeshire be included. Given that this was a national and not a local problem and that all the other dioceses were in competition for the limited intake of new priests and remembering the diminished appeal of a priest's métier when he could not even be guaranteed a living wage, it will be obvious that the task of replacing the dead was prodigious. That it was not altogether impossible is because the Church in 1348, like agriculture, seems to have been a decidedly over-manned profession. Far more of the land-owning families had unloaded their younger sons on the Church than the available parishes could comfortably accommodate and there was therefore a surplus of clergy in search of benefices which could be drawn on at least to plug the more attractive gaps.

But even allowing for this, it was more than the Bishops of Norwich and Ely could do to build up the numbers of their clerics without some loss of standard. Jessop denies that there was any serious decadence in the East Anglian church after the Black Death[35] but certainly the educational level of the new priests was lower than that of their predecessors and, even when they were literate, they were often middle-aged men with little sense of vocation but widowed by the plague and looking for some way to fill the rest of their lives.

Nor were all the new recruits as estimable as this. The Rev. William, priest of the manor of Waltham, was appointed early in 1350. His extra-parochial activities earned him the nickname of

'William the One-day priest,' a title which he amply justified when he held up Matilda, the wife of John Clement de Godychester, and robbed her of her purse. For this offence he was arrested and, one must fear, eventually hanged. The Rev. William was certainly exceptional and such exceptions were not uniquely to be found in the mid-fourteenth century. Delinquent priests cropped up in every generation. But in the conditions of 1350 and 1351 it came as less of a surprise when priests behaved with a striking degree of impropriety. In East Anglia at least, the reputation of the clergy stood lower after the plague than before it.

ELEVEN

The Midlands and the
North of England

The Black Death splattered the central part of England with the same haphazard venom as it had shown in the south. In Huntingdonshire it seems to have followed much the same course as in East Anglia. By 1363, read the preamble to the city charter, Huntingdon 'was so weakened by mortal pestilences and other calamities' that it was quite unable to pay its taxes. A quarter of the town was said to be uninhabited and the remaining residents could scarcely find the means of supporting life.[1] Three churches were derelict, their parishioners either dead or departed. Since the citizens were hoping to get some remission of their taxes it was obviously in their interests to paint the picture as black as possible but, even allowing for this, it is clear that things were in a bad way. But it is worth reiterating yet once again that the misfortunes of Huntingdon, as with many rural areas in England, were not solely due to the Black Death or even to the cumulative effects of the various epidemics. One of the 'other calamities' referred to in the city charter may have been the downfall of one of the local earldoms but more important and more constant was the economic decline of the whole area, and of Huntingdon in particular, which far preceded the violent shock of the plague. The barometer of Huntingdon's health was the success of its great annual fair and the fourteenth century had already provided a dismal history of small attendances and dwindling revenues. The plague, of course, vastly accelerated the process but Huntingdon in 1353 would anyway have found unfairly onerous taxes which it could have paid with little trouble at the beginning of the century.

Northamptonshire seems to have been among the less afflicted of English counties. Mortality among the beneficed clergy of the archdeaconry was just under 37%; a figure which was reasonably constant in all the deaneries. Only in Peterborough, another of those low-lying areas which were so remarkably well treated by the plague, was the level notably below the average at a mere 27%.[2]

Stamford had a disastrous experience. It lost six incumbents in the six months between July and November, 1349, and never recovered the impetus which was carrying it towards the status of an important town. After the Black Death the population remained more or less stable or even continued to decrease; the many references to 'void places' in the deeds of later years suggest that a long time passed before the ravages of the plague had been put right.[3]

Henry Knighton, a canon of Leicester Abbey, who wrote some time after the Black Death but was an eye witness of the disaster, has left an account of the damage done by the plague in the English countryside.[4]

'In this same year,' he recorded, 'a great number of sheep died throughout the whole country, so much so that in one field alone more than five thousand sheep were slain. Their bodies were so corrupted by the plague that neither beast nor bird would touch them. The price of every commodity fell heavily since, because of their fear of death, men seemed to have lost their interest in wealth or in worldly goods. At that time a man could buy for half a mark a horse which formerly had been worth forty shillings. A large, fat ox cost four shillings, a cow one shilling, a bullock sixpence, a fat wether fourpence, a sheep threepence, a lamb twopence, a large pig fivepence and a stone of wool ninepence. Sheep and cattle were left to wander through the fields and among the standing crops since there was no one to drive them off or collect them; for want of people to look after them they died in untold numbers in the hedgerows and ditches all over the country. So few servants and labourers were left that nobody knew where to turn for help.' No such universal or horrifying mortality had taken place within living memory though Bede, in his "De gestis Anglorum" records that, in the time of Vortigern, king of the Britons, not enough people were left alive to carry the dead to their graves.

'The following autumn it was not possible to get a harvester except by paying eightpence a day with food included. Because of this many crops were left to rot in the fields. However, in the year of the pestilence, these crops were so abundant that no one cared whether they were wasted or not.'

In general the countryside was soon back to something near normal. Inevitably, some of the poorer villages were less well able to resist the depredations of the plague and the temptations offered by other, richer neighbours. At Wyville, between Melton Mowbray and Grantham, an inquisition post-mortem found '. . . the three carucates are worth little, for the land is poor and stony, and lies uncultivated for want of tenants after the pestilence.'[5] The village was never fully to recover. But, anyway in the Midlands, such cases were rare; and even Knighton, notoriously gloomy as he was, cou'd have found little to shock him in the outward aspect of Leicestershire by 1354 or 1355.

'The fearful mortality rolled on,' recorded the canon of Leicester, 'following the course of the sun into every part of the kingdom. At Leicester, in the small parish of St Leonard, more than three hundred and eighty people died; in the parish of Holy Cross more than four hundred; in the parish of St Margaret more than seven hundred; and so on in every parish.'[6]

As usual it is impossible to reconcile such figures with what little is known about the total population. Undoubtedly many people died. But the nearest approach to a firmly based estimate suggests that the city could not easily have contained more than three thousand five hundred inhabitants before the Black Death and may even have had less than three thousand.[7] It is difficult to believe that Knighton's figures were more than a picturesque expression of arithmetical inadequacy.

A curious feature is that a surprisingly small drop in the number of tax-payers in Leicester and the amount of tax which they paid was recorded between 1336 and 1354. This, *prima facie*, is hard to reconcile with a death roll that could not possibly have been less than a quarter and was probably a third of the total population. One may permissibly draw several deductions from this. The first and most obvious is that here is another proof of the inaccuracy of Knighton's estimates. Another, slightly more valuable though no more than confirmation of what has already been remarked elsewhere, is that the poor must have borne the brunt of the plague. The tax-payers were naturally the richest section of the community and there is nothing astonishing in the fact that, *pro rata*, many more survived than was the case with their less fortunate fellow-citizens.

But this alone is not enough to account for so sensationally rapid a recovery. The most interesting conclusion to be drawn is that a boom-town like Leicester, centre of a prosperous agricultural area and with rapidly growing trades and industry, could quickly make up its strength recruiting not only peasants but free men of some wealth and standing as well. The same has already been pointed out in the case of other towns but, in Leicester, a local historian has analysed the lists of new inhabitants so as to establish, as nearly as possible, from where they came.[8]

The tallage rolls of Leicester, in the years immediately after the Black Death, record the disappearance of a remarkable number of long-established names and a still more abnormal influx of new ones. Of the two hundred and forty-seven new arrivals who were well enough off to pay taxes by 1354, sixty-five came from villages in Rutland and Leicestershire, another twenty-seven from villages on the borders of Leicestershire and the rest either from unspecified districts or from far afield. Immigrants came from Northumberland, London, Dublin and even Lille; the latter, presumably, to assist with the expanding cloth trade. These figures underline the fact that there was striking mobility among the people of medieval England and that, in the main, the big towns reinforced themselves at the expense of the adjacent countryside. It is unlikely that a village as impoverished as Wyville contributed to Leicester any one rich enough to be a taxpayer within a few years of his arrival but it would not be in the least surprising if some of the few villeins who survived decided that they had had enough of its stony poverty and set off to make their fortune in the city.

In Derbyshire, in the diocese of Lichfield, seventy-seven beneficed clergy died in a hundred and eight parishes. The family of de Wakebridge, records a local historian, lived in considerable style 'in as healthy and uncrowded a spot as any that could be found on all the fair hillsides of Derbyshire.' But their luxury and solitude availed them nothing. Within three months William de Wakebridge lost his father, his wife, his three brothers, his two sisters and a sister-in-law. 'The great plague,' says Dr Cox, 'had the effect of thoroughly unstringing the consciences of many of the survivors and a lamentable outbreak of profligacy was the result.' Sir William, at least, was spared this secondary infection. He retired from the army and gave a large part of his painfully acquired inheritance to the Church.[9]

Nottinghamshire was remarkable mainly for the erratic incidence of the plague within its borders. The mortality among beneficed clergy in the archdeaconry as a whole was 36.5%, well below the national average, but this concealed the difference between Newark, with a calamitous 48%, and Retford, including Sherwood forest and the flat lands in the basin of the Trent, with only 32%. The county as a whole illustrated to a still more marked degree the phenomenon which has already been mentioned in other parts of England; in the high and supposedly healthy county the death rate was far higher than in the flats and fens.[10]

The figures for the clergy in Nottinghamshire illustrate well the danger of stating categorically what in fact is mere assumption. For the one hundred and twenty-six benefices in the county, sixty-five new appointments had to be made. Seebohm assumed that all these were due to the death of the previous incumbent[11] and Gasquet accepted his assumption, concluding '. . . the proportion of deaths among the beneficed clergy is found, as in other cases, to be fully one half the total number.'[12] It was not until Hamilton Thompson, whose figures are quoted above, conducted researches which showed that, in the case of Nottinghamshire, nearly 28% of the benefices were vacated for other reasons, that it became clear how inflated the earlier figures had been.

Neighbouring Lincolnshire suffered more severely than any other county. It was divided into two archdeaconries; Stow, in the north-west, and Lincoln covering the rest of the county. In Stow 57% of beneficed clergy died and in Lincoln 45%, including 57% in Stamford, 60% in the city of Lincoln itself and 56% in the wold deanery of Gartree. A logical explanation for the high figure in Lincolnshire, as suggested in the case of the southern counties, might be that the county's long coast line exposed it to many different lines of attack, in particular by ship-borne rats. But here again logic breaks down. Grimsby, with a death rate of only 35% was one of the two least troubled deaneries in the county; as a prosperous and busy port it should have been at the other end of the scale.[13]

'In 1349,' wrote the common clerk of the city in the Blickling homilies,[14] 'there was that great pestilence in Lincoln which spread all over parts of the world, beginning on Palm Sunday in the year aforesaid and enduring until the feast of the Nativity of St John the Baptist next following [24 June], when it ceased, God be praised who reigns for ever and ever, Amen.'

By 1349, Lincoln was already losing ground economically and some at least of the decline which followed had its origin in earlier causes. But there is not only the evidence of mortality among the parish priests to show how heavily it lost. The Burwarmote Book shows that, of the two hundred and ninety-five wills disposing of burgage tenements which were enrolled in fifty-four years around the year of the plague, a hundred and five can be ascribed to 1349; the fruit, that is to say, of some thirty normal years.[15] Almost all were enrolled in June and July.

The chronicler of Louth Park, the great Cistercian abbey, twenty-five miles north-east of Lincoln, dealt briefly but sufficiently with the outbreak.[16]

'This plague,' he recorded, 'slew Jew, Christian and Saracen alike; it carried off confessor and penitent together. In many places not even a fifth part of the people were left alive. It filled the whole world with terror. So great an epidemic has never been seen nor heard of before this time, for it is believed that even the waters of the flood which happened in the days of Noah did not carry off so vast a multitude. In this year many monks of Louth Park perished; among them, on 12 July, the Abbot, Dom Walter de Luda. He was buried in front of the high altar by the side of Sir Henry Vavasour, Knight.'

No rural economy, however resilient, could recover quickly from devastation on this scale. All generalisations are dangerous but at least it can be said with confidence that the wapentakes at the southern end of the Lincolnshire wolds, 'The classical district of ruined churches and lost village sites,'[17] were left entirely desolate. Centuries were to pass before any serious recolonisation took place. Indeed, it has been convincingly argued that the population of much of the fenlands at the end of the thirteenth century was as large, if not larger, than that recorded in the censuses of the nineteenth century.[18] Fifteen villages in Lincolnshire vanished directly after the Black Death or within a decade or two of its visitation. Probably all of them were thinly populated and economically weak before the middle of the fourteenth century but, in most cases, it must have been the plague which applied the *coup de grâce*.

On 28 July, 1348, Archbishop Zouche of York had taken alarm at the news from the Continent and sent out a warning order to his flock:

'In so far as the life of men upon earth is warfare,' he wrote, 'it is no wonder that those who battle amidst the wickedness of the world are sometimes disturbed by uncertain events; on one occasion favourable, on another adverse. For almighty God sometimes allows those whom he loves to be chastened so that their strength can be made complete by the outpouring of spiritual grace in their time of infirmity. Everybody knows, since the news is now widely spread, what great pestilence, mortality and infection of the air there are in divers parts of the world and which, at this moment, are threatening in particular the land of England. This, surely, must be caused by the sins of men who, made complacent by their prosperity, forget the bounty of the most high Giver.'[19] The Archbishop prescribed the usual course of prayer, processions and litanies to avert the coming of the dreadful pestilence.

When the Archbishop thus addressed his flock, the Black Death had hardly arrived in England. Nearly ten months were to elapse before the first cases were recorded in Yorkshire. For the Englishmen in the south, who had heard with alarm of the terrible plague in Europe, the situation had seemed perilous enough. But France was another country, the Channel lay between; 'it can't happen here' was then, as always, the reaction of the Englishman confronted by disorder among the lesser breeds without the law. But such comfort could not last long. From the moment that the plague got a firm grip in Dorset it must have been clear to every well informed Englishman that, in the end, his turn would come. The more pious or the more optimistic no doubt continued to hope that some miracle would avert their doom but, as the plague moved inexorably northwards, even the most confident must have lost their faith.

Though he may have discounted some of the wilder rumours, any northerner who kept his ears open in the spring of 1349

must have been led to believe that, in all probability, he had no
more than a fifty-fifty chance of surviving the rest of the year. His
family and friends could be expected to die around him and those
in authority over him were more than likely to join him in the
churchyard. Life, as he knew it, seemed on the verge of breaking
down. Faced with such a threat the temptation must have been
great to eat, drink, be merry and do little else besides. It is notable
that, whether through apathy, self-discipline or the resignation
that comes from perfect faith, there seems to have been no such
reaction. Until the moment that the plague reached his village—
with a minor exception in Durham which we shall mention
shortly—even, indeed, when it was already rampant, the average
northerner continued to till his fields, tend his cattle and perform
his manorial duties. The threat of infection was not enough; only
death, it seemed, could distract him from his daily duties.

Though the mortality in Yorkshire was not quite so wholesale
as in other counties, certainly much less so than in neighbouring
Lincolnshire, the Black Death was far from merciful. In the five
hundred and thirty-five parishes of the diocese of York, of which
the great majority were in Yorkshire itself, two hundred and
twenty-three benefices were vacated by death, sixty-three by
resignation and a further fourteen from unspecified causes.
Between 42% and 45% of the parish clergy died.

The huge archdeaconry of York shows the usual wide variants
in the mortality rate from area to area. The deanery of Doncaster
lost nearly 59% of its clergy, yet virtually no benefices were vacated in the marshes between Doncaster and the Humber. In
Pontefract the figure was 40%, again with few or no casualties in
the eastern flats. And yet in the mountainous district of Craven
which, on the analogy of Derbyshire, should have suffered
severely, a mere 27% perished. York itself, with 32%, was relatively lightly touched.

In so vast a county it will be obvious that the Black Death
could not have arrived everywhere at the same time. It is most
unlikely that more than a handful of cases occurred before April,
1349; Pope Clement VI's Bull of 23 March from Avignon referred to the plague as having already begun to harass the diocese
but it seems certain that this was premature and that, even if there
had been an outbreak in March or early April, it was confined to
Nottinghamshire or perhaps to one or two coastal areas. It did

not arrive in the city of York until 21 May. June, July and August were the worst months all over the north of England.

The Archbishop of York held to his normal practice and discreetly sat out the summer months at one or other of his rural manors near the little town of Ripon.[20] His suffragan, however, Archbishop Hugh of Damascus, behaved with a vigour unusual in a senior churchman. He was an Austin friar who had once been excommunicated by Archbishop Grandisson for outrageous behaviour.[21] This unconventional background perhaps explained why he so far forgot his rank as to tour his diocese, visit the sick, encourage the healthy and consecrate many new churchyards. These were usually only authorised for temporary use and the citizens of Fulford, a mile or so south of York, found themselves in trouble some years later when they persisted in burying their dead in their new churchyard of St Oswald long after the Black Death had passed. Archbishop Thoresby, in a curious commentary on the pleasures of the times, accused them of 'taking an empty delight in novelty' and ordered them to resort forthwith to their traditional burial ground.[22]

York was one of the largest cities in England; smaller only than London and, perhaps, Norwich. Professor Russell[23] estimates the population in 1377 at ten thousand eight hundred and seventy-two. By that date economic expansion had already been resumed and the plague, destructive though it was, provided no more than a check in a process which lasted till the end of the century.[24] But, at the time of the Black Death, the city was already in difficulties, since a disastrous flood had submerged its western parishes on 31 December, 1348.[25] Jeanselme, who contends that floods, famines and earthquakes invariably precede an epidemic of plague, indeed cites the flooding of the Ouse as evidence of his thesis.[26] It is hard to see why, in that case, the plague should have spread with equal vigour to all the other cities where the Ouse did not overflow and no earthquake or famine took place. But one can accept that the commotion which the sudden flood must have caused among the rats of York could have helped the spread of infection throughout the city and the neighbourhood.

At the Abbey of Meaux, six miles north of Hull, the monks were already expecting the worst. On the Friday before Passion Sunday, they recalled, they had been violently thrown from their stalls by an earthquake. Such a portent could only have meant trouble to come. When it was coupled with the undoubtedly sinister birth of Siamese twins in Kingston-upon-Hull and their death a few days before, then even the most sceptical could hardly have doubted the imminence of disaster. Disaster duly came. '. . . God's providence ordained at that time,' wrote the chronicler of the monastery, 'that in many places the chaplains were kept alive to the very end of the pestilence in order to bury the dead; but after this burial of the lay folk the chaplains themselves were devoured by the plague, as the others had been before them.'[27] The Abbot, Hugh de Leven, and five monks died in a single day, on 12 August, 1349; the prior, the cellarer, the bursar and seventeen other monks died also and, out of fifty monks and lay brethren, the chronicler claims that only ten survived.[28] The Abbey found its revenues so reduced that it was forced to lease its grange for life to a certain Sir William de Swine for a lump payment of £80. Whether by Divine intervention, coincidence or monkish initiative, Sir William was murdered shortly afterwards and the grange reverted to the Abbey. But, in spite of this windfall, recovery was slow and, in 1354, the Abbey was handed over to a royal commission 'on account of its miserable condition.'[29]

The countryside around Hull seems to have been ravaged simultaneously by plague and by the flooding of the Rivers Hull and Humber. An inquisition post-mortem at Hemingbrough showed pasture normally worth 12d trampled down and valueless: 'The herdsmen were lying in the churchyard.'[30] All the bondigers and carters of the Abbey of Selby were dead so that turves and hay had to be carried by water. As for Hull itself, the King in 1353 remitted certain taxes: '. . . considering the waste and destruction which our town of Kingston-on-Hull has suffered, both through the overflow of the waters of the Humber and other causes, and that a great part of the people of the said town have died in the last deadly pestilence which raged in these parts and that the remnant left in the town are so desolate and poverty-stricken.'[31] At Wharram Percy the population of the parish was so reduced that, shortly after the Black Death, the side aisles of the church were pulled down and never rebuilt.

But in Yorkshire as elsewhere the structure of government was strained but substantially survived. A curious illustration of this comes from the city of York itself. On 7 August, 1349, a jury

certified that William Needler had died 'a natural and not violent
death by reason of the pestilence in Coppergate, York.' There
must have been some decidedly suspicious circumstances about
his death if, when so many were perishing, it was found necessary
to conduct an inquest. But the fact that it was possible to assemble
a jury and record a verdict according to the due processes of the
law at a time when the epidemic in the city must have been at or
near its peak is an impressive tribute to the civic sense of the
inhabitants or, perhaps, to the discipline with which the authorities imposed their rule.

Lancashire was at this date one of the most thinly populated of
the English counties. One area about which an unusual amount
of information exists is the deanery of Amounderness to the west
of the county, including the parishes of Preston, Lancaster and
Blackpool. The Archdeacon of Richmond and his proctor, the
Dean of Amounderness, had a dispute about certain fees for the
probate of wills. A jury was set up to inquire into the question.
They viewed the statistics presented to them with a certain amount
of scepticism but no alternative totals were suggested and figures
of this kind at least carry slightly more conviction than the vague
calculations of the chronicles. In the ten parishes of Amounderness it was claimed, thirteen thousand one hundred and eighty
people died between 8 September, 1349 and 11 January, 1350. In
the parish of Preston, three thousand died, in Poulton-le-Fylde,
it was said, eight hundred died; in Lancaster, three thousand;
Garstang, two thousand and Kirkham, three thousand.[32]

Even allowing for the enormous size of the parishes and some
inflation in the estimates of the dead it seems that Amounderness
must have suffered worse than the regional average. But evidence
is scarce. At Rochdale, where the parson died, there was a gap of
eight months before the vacancy was filled, but this could be
accounted for by inefficiency on the part of the diocesan authorities as well as by the high mortality. A curious entry is found in
the Assize Rolls. A certain William of Liverpool 'caused one
third of the inhabitants of Everton to be brought to his house
after death'; presumably with the intention of carrying out cut-
price funerals and so cheating the lord of the manor and the
church authorities. Since it was widely assumed that the plague

was caught from the dead, his courage deserves praise as well as his business acumen.[33]

Of Cumberland still less is known. In this county, the diocesan registers are lacking; a tribute not so much to the Black Death as to the havoc wrought by the invading Scots. But though the Scots prepared the ground it was the plague which finally dislocated the agricultural economy. The accounts of Richard de Denton, former Vice-Sheriff, presented for audit in 1354, show vividly what damage had been done. Because of 'the mortal pestilence lately raging in those parts,' he reported, 'the greater parts of the manor lands attached to the King's Castle at Carlisle' were still lying uncultivated. For eighteen months after the end of the plague, indeed, the entire estate had been let go to waste 'for lack of labourers and divers tenants. Mills, fishing, pastures and meadow lands could not be let during that time for want of tenants willing to take the farms of those who died in the said plague.' The jury found that Richard de Denton had proved his facts and accepted the greatly reduced value of the estate.[34] The city of Carlisle was relieved of many of its taxes in 1352 because 'it is rendered void and, more than usual, is depressed by the mortal pestilence.'

Durham too had suffered severely from the incursions of marauding Scots. In 1346 they had invaded in greater numbers than ever before. Under the sacred banner of St Cuthbert, Bishop Hatfield had taken the field and repelled them. But, though the victory was decisive, it had not come in time to save the Palatinate from devastation. Against such a background it is not surprising that the morale of the inhabitants should have been frail even before the threat of the Black Death became imminent. Durham is almost the only county in England where there is any evidence of panic spreading before the arrival of the epidemic and it is reasonable to see a link between this and the recent tribulations of the area. Yet even here no very dramatic evidence of demoralisation is to be found.

That summer the halmote at Chester le Street opened as usual but when the Bishop's steward arrived on 15 July at Houghton le Spring he found that accounts of the plague had spread dismay among the peasants. 'There was no one,' it was recorded, 'who would pay the fine for any land which was in the lord's hands through fear of the plague.' At Easington, the next centre for the halmote, things were even worse. The steward offered to make payment of rent contingent on the tenant's survival of the plague but even this could not tempt the nervous peasants into taking on any new responsibility. In the end he was forced to let three tenements at an absurdly low rent since even this would be of greater benefit to the lord than to leave the land untilled.[35]

In his history of Durham, Surtees[36] described the Scottish invasion and concluded, 'No other events than those related disturbed the peace of Hatfield's Pontificate.' The point of view which could thus lightly dismiss a calamity which killed perhaps ten times as many people as the battles with the Scots is hard to understand. But in justice to Surtees it must be admitted that few details are known. The usual pin-points of light illumine the great obscurity. Billingham was badly affected; forty-eight of the prior's tenants were carried off, probably well over half the population. A laconic entry in the Bishop's rolls records, 'No tenant came from West Thickley because they are all dead.' A peasant, driven mad with grief by the loss of all his family, wandered in search of them from village to village of the Palatinate. For many years his unceasing quest was to revive ugly memories throughout the countryside.[37]

It seems that in Durham relations between landlord and tenant suffered exceptionally as a consequence of the plague. Here too the damage done by the incursions of the Scots may have contributed to the malaise. At all events, while the Black Death was waning, something close to a strike took place in several villages of the Chester ward and harsh methods of repression had to be adopted.[41] Too little information survives to give any real idea whether this was no more than a spasm of resentment against an unpopular bailiff or a wider and more serious movement against authority; it is at all events curious that the last county in England to be visited by the plague should have been the first to yield any evidence of rural disorder.

TWELVE

The Welsh Borders, Wales, Ireland and Scotland

But while the Black Death had thus moved northwards to the Scottish border, Wales and the adjoining British counties had not been spared. From Bristol the plague had spread into Worcestershire, rising to its crescendo in June, 1349; then dying away in August only to return in the late autumn.[1] As early as April it had proved necessary to forbid further burials in the cathedral churchyard at Worcester because the congestion of the dead was beginning to threaten the survival of the living.[2]

'Alas,' recorded the Bishop, 'the burials have in these days, to our sorrow, increased . . . (for the great number of the dead in our days has never been equalled); and, on this account, both for our brethren in the said church ministering devoutly to God and His most glorious Mother, for the citizens of the said city and others dwelling therein, and for all others coming to the place, because of the various dangers which may probably await them from the corruption of the bodies, we desire, as far as God shall grant us, to provide the best remedy.'[3]

The Bishop's remedy was to open a new graveyard beside the hospital of St Oswald and transfer there not only all burials which would otherwise have been in the Cathedral cemetery, but also from several of the parish churches of Worcester as well. 'Hence,' in the lapidary phrase of a local historian, 'that prodigious assemblage of tumulation which, at this time, cannot be viewed with indifference by the most cursory beholder.'[4]

Bishop Wulstan Bransford himself remained secluded in his manor at Hartlebury, four miles south of Kidderminster. In spite of this precaution he died on 6 August, 1349. The King's Escheator reported on the state of his estates between early August when he died and late November when a successor was appointed. His record shows that Hartlebury was not an isolated case. Tenants, he said, could not be got at any price; mills were vacant, forges standing idle, pigeon houses in ruins with all the birds fled. Of £140 owing to the Bishop in cash or in the form of various feudal services, £84 were never received, '. . . on account of the dearth of tenants, who were wont to pay rent, and of customary tenants, who used to perform the said works, but who all died in the deadly pestilence.' As late as 1354, relief was still being sought on the grounds that it was impossible for the Bishop to obtain any of the customary services which had once been his due; '. . . the remnant of the said tenants had changed them into other services and, after the plague, they were no longer bound to perform services of this kind.'

Some time in 1349 a serious riot took place between the townsmen and the monks of the Priory of St Mary, the Cathedral monastery. The townsmen broke down the gates of the priory, chased the prior 'with bows and arrows and other offensive weapons' and tried to set fire to the buildings. Here, as in the somewhat similar incident at Yeovil,[5] it is tempting to see some link with the Black Death. Certainly such a possibility cannot be excluded. But chasing the prior with bows and arrows and other offensive weapons was by no means unheard of in Worcester. Relations between town and cathedral monastery were often strained in medieval England and, though the Black Death may have heightened the tension, there is no reason to believe that in Worcester or elsewhere it actually created it.

Bishop Trilleck's neighbouring diocese fared no better. In Hereford the Bishop forbade the acting of 'theatrical plays and interludes' in the city churches, a belated attempt to avert the wrath of the Almighty which seems to have met with little success. In the end, it is claimed, the epidemic was checked 'by carrying the shrine of St Thomas of Cantilupe in procession.'[6] In 1352 a joint petition was lodged by the patrons of the two churches of Great Colington and Little Colington:

'the sore calamity of pestilence of men lately passed, which ravaged the whole world in every part, has so reduced the number of the people of the said churches and for that said reason there followed, and still exists, such a paucity of labourers and other inhabitants, such manifest sterility of the lands, and such notorious poverty in the said parishes, that the parishioners and receipts of both churches scarcely suffice to support one priest.'

The Bishop saw the justice of the complaint and the two parishes were duly amalgamated.

The county historians illustrate the impact of the plague by quoting the inquisition post-mortem on the family of John le Strange of Whitchurch.[7] John died on 20 August, when the plague had already done its worst over most of the county. He left three sons: Fulk, Humphrey and John the younger, of whom Fulk, as eldest, was naturally the heir. By the time the inquiry was held on 30 August, Fulk had already been dead two days. Before an inquisition could be held on Fulk's estate, Humphrey too was dead. John, the third brother, survived but inherited a shattered estate. Even before his father died the three water mills 'which used to be worth twenty marks' had been reassessed at only half the value, 'by reason of the want of those grinding, on account of the pestilence.' In another of his manors, 'two carucates of land which used to be worth yearly sixty shillings' were held to be worth nothing 'because the domestic servants and labourers are dead and no one is willing to hire the land.'

Cheshire, to the north, was thinly populated in the fourteenth century but there is plenty of evidence to show that the losses were still severe. The heads of three of the largest religious houses—the abbot of St Werburgh's, the prioress of St Mary's, Chester and the prior of Norton—all died within a few weeks of each other. It was impossible to find anyone able and willing to hold the eyre of the forest, the bridge over the Dee remained out of repair for several months, the income gained from tolls at the passage of Lawton dropped away to little more than half its former value. The increased bargaining power which the Black Death put into the hands of the surviving tenants is well illustrated by the case of the manor of Rudheath, between Northwich and Macclesfield. A note on the Court Roll reads:[8]

'In money remitted to the tenants . . . by the Justices of Chester and others, by the advice of the Lord, for the third part of their rent, by reason of the plague which had been raging, because the tenants there wished to depart and leave the holdings on the Lord's hands unless they obtained this remission until the world do come better again, and the holdings possess a greater value . . . £10 13s 11¾d.'

It would be interesting to know more about the status of the tenants. The Government was shortly to pass legislation seeking to prevent the migration even of free tenants but it would not be surprising if the tenants of Rudheath in fact enjoyed no legal right to quit their tenements and were blackmailing their land-lord with a threat to commit an unlawful act. Certainly such a case would not have been unique. The knowledge that the law was on his side was small comfort to a lord whose tenants had escaped and were now working on some neighbouring estate, enjoying more favourable terms and sheltered by their new master who, however he might deplore their breach of the feudal laws, was still primarily interested in ensuring that his own houses were lived in and his own lands were tilled.

Mr Rees, the leading authority on the Black Death in Wales, has recorded the lament of the contemporary Welsh poet, Jeuan Gethin, who must have seen and described the plague in March or April of 1349:[9]

'We see death coming into our midst like black smoke, a plague which cuts off the young, a rootless phantom which has no mercy for fair countenance. Woe is me of the shilling in the arm-pit; it is seething, terrible, wherever it may come, a head that gives pain and causes a loud cry, a burden carried under the arms, a painful angry knob, a white lump. It is of the form of an apple, like the head of an onion, a small boil that spares no one. Great is its seething, like a burning cinder, a grievous thing of an ashy colour. It is an ugly eruption that comes with unseemly haste. They are similar to the seeds of the black peas, broken fragments of brittle sea-coal and crowds precede the end. It is a grievous ornament that breaks out in a rash. They are like a shower of peas, the early ornaments of black death, cinders of the peelings of the cockle weed, a mixed multitude, a black plague like halfpence, like berries. It is a grievous thing that they should be on a fair skin.'

'Black smoke,' 'Rootless phantom': such phrases convey some-thing of the mystery and horror of the plague to those who suffered it. But what is so moving about Gethin's comment is that he did not allow the horror to overwhelm him but, with the vocabulary of a poet and the eye of a scientific observer, struggled to pin down the physical appearance of the phenomenon, to find the simile which would convey to the reader precisely what he saw. It was the defiant dedication of the doctor who, on his deathbed, records his symptoms from moment to moment for the future education of his colleagues. In the response of men like Jeuan Gethin lay the victory of mankind over his adversities.

Geoffrey the Baker traced the course of the plague around England. 'The following year,' he went on, 'it devastated Wales as well as England . . .'[10] The chronicler's phrasing is obscure but it seems from the context that 'The following year' must refer to 1350. This must be incorrect. The Welsh were affected at much the same time as their English neighbours; in the south, indeed, somewhat earlier since the infection apparently moved across the Severn valley into Monmouthshire before it had run its course through Gloucestershire and Worcestershire.[11]

By March, 1349, the infection had taken a firm grip on the whole lordship of Abergavenny. The lord of the eastern portion died at the beginning of the month, and by the middle of April, devastation was almost complete. In the manor of Penros only £4 out of rents worth £12 could be collected 'because many of the tenements lie empty and derelict for lack of tenants.' The guardian of the heir petitioned for a reduction of £140 in a rent of £340. An inquiry allowed arrears only to the extent of £60 but, more significantly, accepted a permanent reduction of £40. The damage must have been grave indeed if the sceptical royal officials were prepared to accept that there was no hope of it being made good in the foreseeable future.

So far as any course can be plotted the disease seems to have travelled northwards through the border counties of Hereford, Shropshire and Cheshire and re-entered Wales in the North-East. The lead miners at Holywell, a few miles west of Flint, suffered so severely that the survivors refused to go on working. The Court Rolls of Ruthin provide an unusually complete picture of the depredations of the plague in that part of Wales. Nothing at all unusual seems to have happened before the end of May. Then, in the second week of June, the abnormal number of seven deaths took place within the jurisdiction of the Court of Abergwiller. The plague quickly spread. Seventy-seven of the inhabitants of Ruthin died within the next two weeks; ten in Llangollen, thirteen in Llanerch, twenty-five in Dogfeiling. Mortality continued at this level or even higher until the middle of July, abated for a few weeks, then returned to its most ferocious excesses in the last three weeks of August. The worst was then over and the winter passed with relatively little further loss.

Rees considers that the Black Death probably reached Carmarthen by way of the sea. Certainly two of the officials of the Staple were among the first victims and, if infected boats were putting into the harbour, their post would have been one of peculiar danger. The Lord of Carmarthen, in fact the Prince of Wales, suffered no less than other great landlords: receipts from mills and fisheries fell drastically and fairs, one of the most profitable sources of revenue, had to be abandoned altogether.

In Cardigan, so great was the mortality and the fear of infection that it proved almost impossible to find anyone to fill such offices as beadle, reeve, or serjeant. Out of one hundred and four gabularii or rent-paying tenants, ninety-seven died or fled before midsummer.

Wales in the mid-fourteenth century was divided into the lowland 'Englishry,' largely controlled by colonisers from across the border and run on a manorial basis similar to that of England and the upland 'Welshry' where the unfortunate natives skulked in what was left to them of their country. In the latter areas the writ of the English hardly ran and such records as survive give little indication of what befell the inhabitants. That they suffered seems certain and, if the analogy of the English hills is anything to go by, they suffered worse than their invaders in the valleys. But the damp mist which hangs so constantly over the Welsh mountains seems as apt to confound the historian as the tourist and even the small nugget of fact on which large guesses can be based is here entirely lacking.

Painful readjustment, demoralisation, lawlessness: such are the familiar symptoms of a society recovering from the shock of the plague. Madoc Ap Ririd and his brother Kenwric 'came by night in the Pestilence to the house of Aylmar after the death of the wife of Aylmar and took from the same house one water pitcher and basin, value one shilling, old iron, value fourpence. And they also present that Madoc and Kenwric came by night to the house of Almar in the vill of Rewe in the Pestilence, and from that house stole three oxen of John le Parker and three cows, value six shillings.'[12] How many others must have 'come by night in the

Pestilence,' to profit by the concomitant chaos, to rob the sur-vivors or loot the houses of the dead.

But in Wales as in England, though law and order was badly shaken, substantially it survived. Burglary and banditry were anyhow far from uncommon in medieval England and self-defence the only satisfactory answer to the would-be aggressor. Things certainly got worse at the time of the Black Death but not sensationally so. The main highways were little less safe than in the past; the streets in the cities and big towns, anyhow never to be recommended during the hours of darkness, do not seem to have become conspicuously more perilous. In some cases, where the authorities lost their grip, the more prosperous citizens formed vigilance committees and took their protection into their own hands. A great many Aylmars were fated to lose, not only their wife but their pitcher and their old iron, value fourpence, as well. But the situation never became intolerable. Certainly the greater lawlessness was an inconsiderable extra burden compared with the overwhelming weight of the plague itself.

Mr Rees records that the effects of the Black Death in Wales seem to have been very similar, at least so far as the Englishry was concerned, to the effects in England itself. The decay of the manor and the manorial system was the immediate and the per-manent consequence of the plague. The garden of the manor, with no one to tend it, was more and more often let out as pasture. The dovecote and fish stew were allowed to fall into disuse and often never reactivated. The lords of the manor renounced the farming of the manorial demesne and began to let it out at the best rent they could get. The principle of bondage thenceforward played a far less significant part in the social structure of the manor. The system, in short, broke down because of the shortage of labour and the improved bargaining position of the villein.

All these phenomena were recorded in England too. But in the latter country so many qualifications have to be made to allow for the history of the previous decades, for regional variations and for eccentric and inexplicable movements against the trend that any generalisation is open to destructive criticism. In Wales the scope for generalisation is greater. Partly the reason for this is geographical: the area was smaller and more homogeneous; variations therefore were less. But the nature of the manorial system in Wales ensured that it would bear the imprint of the plague in a way much more clear-cut and decisive than its English counterpart. On the one hand the seeds of decay, which were already beginning to corrupt the English system long before *Pasteurella Pestis* added its contribution, had by 1349 hardly affected Wales. Any change which did take place at this period can therefore be attributed with greater confidence to the plague. On the other hand, since the manorial system in Wales was younger and more fragile, it succumbed more rapidly to the blows which it received in 1349. In Wales the Black Death accomplished in a year or two a revolution which in England was worked out over the whole of the fourteenth century.

In part this statement depends for its validity upon a comfortable foundation of ignorance. Very little is known about the Black Death in Wales and far less work has been done upon the evolution of the manorial system there than is the case with its English parent. No doubt a greater knowledge of the facts would suggest the need for important qualifications. But it is unlikely that the central proposition would be overthrown. The generalisation so often made and so often disputed in the case of England—that the Black Death was directly responsible for the ending of the manorial system—can with greater confidence be applied to Wales.

But even here one is on shaky ground. For before the effects of the Black Death had fully worked themselves out, a cataclysm in some ways still more violent had fallen on Wales. The wars of independence of Owen Glendower, however noble or well-justified, set back the economic and social development of Wales by two hundred years. Through the thick clouds of hatred and bloodshed, through the appalling destruction and loss of life, it is difficult to see clearly what lay before and impossible to deduce how things would have developed but for the obliterating catastrophe. That the Black Death altered Wales is certain but the dimensions of the change can be no more than speculation.

'And I, Brother John Clyn, of the Order of Friars Minor and of the convent of Kilkenny, wrote in this book those notable things which happened in my time, which I saw with my own eyes, or which I learned from people worthy of belief. And in case things which should be remembered perish with time and vanish from the memory of those who are to come after us, I, seeing so many evils and the whole world, as it were, placed within the grasp of the evil one, being myself as if among the dead, waiting for death to visit me, have put into writing truthfully all the things that I have heard. And, lest the writing should perish with the writer and the work fail with the labourer, I leave parchment to continue this work, if perchance any man survive and any of the race of Adam escape this pestilence and carry on the work which I have begun.'[13]

John Clyn added two words to his peroration: *magna karistia* —'great dearth' Then he joined his fellows; another hand briefly added at some later date, 'Here it seems that the author died.'

Even if no other evidence survived from Ireland, John Clyn's cry would show how painfully the country must have suffered. He was a lonely, frightened man, who had already witnessed the death-agonies of almost all the other members of his house and now sought to record their end for posterity before the oblivion of death swept over all Kilkenny and all the country—even all the world. Whether anyone would live to read his words he did not know, hardly dared even wonder, but that instinct which leads men to seek to communicate with their unknown successors, whoever they might be and whatever they might be doing, now drove him on to write his chronicle, a memorial to the terror and grief of those who were still alive.

There is still much that is obscure about the course of the Black Death in Ireland.[14] We cannot even be sure from whence it came. The most likely source is Bristol which was then the main centre of Anglo-Irish trade, but it could well have come direct from Gascony or one of the ports of Brittany. More important and considerably more mysterious is the period of the epidemic. John Clyn was categoric. Referring to 1348 he said:

'. . . in the months of September and October, bishops, prelates, priests, friars, noblemen and others, women as well as men, came in great numbers from every part of Ireland to the pilgrimage centre of That Molyngis. [Teach Molinge on the River Burrow.[15]] So great were their numbers that on many days it was possible to see thousands of people flocking there; some through devotion but others (the majority indeed) through fear of the plague, which then was very prevalent. It began near Dublin at Howth and at Drogheda. These cities were almost entirely de-

stroyed and emptied of inhabitants so that in Dublin alone, between the beginning of August and Christmas, fourteen thousand
people died.'

There is no more reason to take Clyn's statistics seriously than
those of any other chronicler but, equally, there is no reason to
expect him to be seriously wrong over dates. On this basis, therefore, Ireland must have been infected within a month or two of
England. On the whole a rather longer delay was to be expected
but there is nothing wildly improbable in such a conclusion.

Yet in August, 1349, Richard Fitzralph, Archbishop of Ar-
magh, told the Pope during a visit to Avignon that the plague
had destroyed two thirds of the English nation but had not yet
done any conspicuous harm to the Irish or the Scots.

Such minor outbreaks as there had been were confined to the
coastal areas. Fitzralph may have been a few weeks behind the
times with his information but he went directly from Ireland to
Avignon and would anyhow have kept closely in touch with
affairs in his diocese. It took a minimum of fifteen or sixteen days
to get from London to Avignon and an allowance of four weeks
was not considered over-generous. From Dublin it would probably have taken a few days longer. But on any calculation the
Archbishop must have been aware of any major calamity in
Ireland which happened before the end of June. Even allowing
for hyperbole on the part of John Clyn, it is incredible that the
Archbishop should have dismissed as of minor importance an
epidemic which could be described as having 'almost entirely
destroyed and emptied of inhabitants,' Howth, Drogheda and
Dublin.

Other information supports Richard Fitzralph. The Archbishop of Dublin died on 14 July, 1349, and the Bishop of Meath
in the same month. If Fitzralph had left Ireland about the middle
of July it is not at all surprising that he should not have had this
news by the time of his audience with the Pope. But it is more
surprising that both deaths should have occurred almost a year
after John Clyn's plague had ravaged Ireland. Further evidence
from the Annals of Connacht for 1349 lead to the same conclusion. 'A great plague in Moylurg' these record 'and all Ireland
this year. Matha, son of Cathal O'Ruairc died of the plague. The
Earl's grandson died. Risdered O'Raigillig, King of East Brefne
died.'[16] Clyn may have expressed himself badly and meant that,

though the first cases of the plague were recorded in 1348, the epidemic did not become serious until 1349, in particular the summer and autumn. For the want of a better explanation this will have to suffice. Certainly John Clyn can be excused a certain stylistic looseness given the circumstances in which he wrote.

Whatever the dates of the epidemic there is ample evidence of its disastrous impact. Even while Fitzralph was making his way to Avignon it was spreading out from the Pale on the east coast to the midlands and the west. John Clyn records that, of his own Friars Minor, twenty-five died in their house at Drogheda and twenty-three at Dublin. In July, 1350, the Mayor and Bailiffs of Cork filed a petition pleading for the revision of certain taxes. Clonmel and New Ross also petitioned successfully for relief. The citizens of Dublin begged for a special allowance of a thousand quarters of corn. As late as 1354, the tenants of certain royal farms around the capital were claiming that they had been reduced to pauperdom by the 'plague lately existing in the said country,' and because of 'the excessive price of provisions' which was exacted by certain royal officials. Geoffrey the Baker states that the Anglo-Irish were almost wiped out but the pure blooded Irish in the mountains remained inviolate till 1357.[17] This cannot be accepted but it is possible that comparatively little damage was done among the indigenous Irish and that these suffered more severely in another epidemic eight years later—perhaps of some quite different disease.

But to see the sufferings of Ireland in their proper perspective it is necessary to remember that, in appealing for relief, reference was usually made not only to the plague but also to the 'other many misfortunes which had happened there'—'the destruction and wasting of lands, houses and possessions by our Irish enemies.' An inquisition of the lands of Roger de Mortimer found 'a great and flourishing manor, full of free tenants, farmers and burgesses, waste'. The manor of Geashil, belonging to the Earl of Kildare, was 'worth nothing.' The demesne lands in County Longford 'lay waste for lack of tenants.' All these statements of sad fact sound familiar enough and could have been culled from the records of any county of England which was recovering from the plague. But in Ireland they stem from the 1320s and 1330s; fruit not of the plague but of the perpetual, ruinous civil war which ravaged the country in the fourteenth as in almost every other century. The Black Death was no less painful to the Irish because they were accustomed to live in a state of even bloodier disorder than their neighbours across the Irish Sea but it should not be forgotten that a high proportion of their misfortunes would have arisen even though there had been no plague to help them forward.

It would be proper to conclude this tour of the Black Death in Britain with some account of its spread to Scotland. Unfortunately, little detail is recorded. According to Knighton[18] the Scots were delighted when they heard of the fate which had overtaken their hated neighbours in England. They regarded it as a proper retribution for past offences and, as the plague swirled over Cumberland and Durham, massed their forces in the forest of Selkirk, 'laughing at their enemies' and awaiting the best moment to invade. It was their last laugh for, as they were on the point of moving into action, 'the fearful mortality fell upon them and the Scots were scattered by sudden and savage death so that, within a short period, some five thousand died'. The panic-stricken soldiers dispersed throughout Scotland, dying by the side of the road or carrying the infection with them to their homes.

'God and Sen Mungo, Sen Ninian and Seynt Andrew scheld us this day and ilka day fro Goddis grace and the foule deth that Ynglessh men dyene upon.'[19] Such was the prayer that the Scottish soldier was trained to say as he watched the sufferings of the English on the other side of the frontier. It is certain that his prayers went for little but so little attention is paid to the Black Death by historians of Scotland that one is tempted to believe that St Mungo, St Ninian and St Andrew must have spared their admirers some part at least of the tribulations of other less well protected Europeans. One of the few people living at the time of the plague whose account survives is John of Fordun.[20]

'In the year 1350,' he wrote, 'there was, in the kingdom of Scotland, so great a pestilence and plague among men . . . as, from the beginning of the world even unto modern times, had never been heard of by man, nor is found in books, for the enlightenment of those who come after. For, to such a pitch did that plague wreak its cruel spite, that nearly a third of mankind were thereby made to pay the debt of nature. Moreover, by God's will, this evil led to a strange and unwonted kind of death, insomuch that the flesh of the sick was somehow puffed out and swollen, and they dragged out their earthly life for barely two days. Now this everywhere attacked especially the meaner sort and common people;—seldom the magnates. Men shrank from it so much that, through fear of contagion, sons, fleeing as from the face of leprosy or from an adder, durst not go and see their parents in the throes of death.'

This all-too familiar description could have applied as well to any other country. Androw of Wyntoun,[21] who was contemporaneous with John of Fordun but certainly never read the latter's chronicles, confirms that Scotland suffered severely.[22]

> 'In Scotland, the fyrst Pestilens
> Begouth, off sa gret wyolens,
> That it was sayd, off lywänd men
> The thyrd part it dystroyid then
> Efftyr that in till Scotland
> A yhere or more it was wedand
> Before that tyme was nevyr sene
> A pestilens in oure land sa kene:
> Bathe men and barnys and women
> It sparryed noucht for to kille them.'

Finally, the Book of Pluscarden,[23] a slightly later chronicle but still written close enough to the date of the plague to have some ring of authenticity, also refers to a third of the population being slain and to the poor suffering far worse than the rich.

'They were attacked with inflammation and lingered barely four and twenty hours,' noted the anonymous author, concluding more hopefully: 'The sovereign remedy is to pay vows to St Sebastian, as appears more clearly in the legend of his life.'

The most striking feature of these accounts is the reiterated statement that a third of the population perished. This is a conservative figure compared with the speculation of the chroniclers of other countries whose estimates of the mortality might be anything between fifty and ninety per cent, or even on occasion the entire population. It is perhaps to be expected that the medieval statisticians of Scotland would approach their task with a sobriety not to be found in the English or the still more volatile Latins but, even so, it seems unlikely that they can be acquitted

entirely of exaggeration. Unless their estimates were very far out
of line with those of other countries the figure of a third must
have been substantially too high. If this is so, then Scotland must
have escaped more lightly than England or Wales.

The emphasis given to the virtual immunity of the rich and
powerful is also interesting. Everywhere it was the poor who
suffered worst and, generally speaking, the more eminent the
individual's position, the greater his chances of survival. To take
only one example: 18% of the English Bishops died as against
some 40% of all beneficed clergy. But it was by no means unusual for the great to perish; there are innumerable cases of
noblemen or merchants, living in large and spacious houses, who
met the same fate as their less prosperous neighbours. Clearly
such cases cannot have been unknown in Scotland but John of
Fordun's emphatic statement that 'the meaner sort and common
people' were above all the victims suggests that the discrepancy
was even more marked north of the Tweed.

With lower overall mortality and relatively trivial losses among
the nobility and upper levels of society it is less surprising that
the plague should have left so light a scar in Scotland. It does not
seem, however, to have disturbed the balance of power between
England and Scotland, though the failure of the English to win
the lasting success which had seemed to be made possible by the
rout of the Scots at Neville's Cross and the capture of King David,
can perhaps in part be blamed on the shortage of man-power
which resulted from the plague. At all events, the appetite of the
Scots for plunder and revenge was temporarily checked and it
was several years before they recovered their full zest for forays
across the border.

Though cases of the plague occurred north of the border in the
autumn of 1349, it seems to have been largely held in check by
the Scottish winter. This can, to some extent, be ascribed to the
reluctance of rats to change their residence in intense cold but
winter does not seem to have been much of an impediment to the
advance of the Black Death in other countries. Whatever the
reason, the lull was short-lived. In the spring of 1350, the plague
was on the move again and quickly blanketed the whole country.
By the end of that year, all Britain had been infected; all Europe,
indeed, since the countries to the north were ravaged at much
the same time. In the whole continent, with the exception of a few

lucky pockets, hardly a village was left unscathed, hardly an individual can have escaped without the loss of at least one friend or relative. It was a continent in mourning. Millions of fresh graves provided a visible memorial but it was not only the dead who paid the price. To survive the Black Death was not to survive unscathed. Indeed, in some ways, the shock which it inflicted on the minds of men seemed even more significant than the fearful harvest which it had reaped among their bodies.

THIRTEEN

The Plague in a
Medieval Village

Statistics alone cannot provide an adequate picture of the Black Death. That 48.6% of the beneficed clergy in a given diocese died between April and September, 1349, is an imposing but somewhat flavourless concept which, in itself, gives no very vivid impression of the sufferings of the people. That a quarter, a third or even half the population died as well is more striking, but the figures still convey no proper idea of what so brutal a depopulation meant to those who survived. In every country the great majority of those who lived and those who died were village dwellers, dependent on agriculture for their existence.

The academic historian rightly distrusts, even if he does not despise, the work of imaginative reconstruction produced by the historical novelist. *A fortiori*, there must be excellent reason to justify the introduction into a book of this kind of any detail which lacks some sort of documentary evidence. But if the effect of the Black Death is really to be understood then it must be studied at work in a small village community and some attempt be made to evoke the atmosphere which it created and which it left behind it. Not enough is known about any one village to make this possible, but, by piecing together scraps of authenticated material, it is possible to construct a coherent picture which, in essence, is plausible and valid. Only by such an exercise can one hope to put flesh on the dry statistical bones provided by the records of the period.

The village of Blakwater, then, is imaginary; that is to say it is not to be found on any map and was unknown to the compilers of the Domesday Survey. But in its organisation and its composition it is not in the least a work of imagination; on the contrary it is very ordinary, and every feature could be duplicated in many hundreds of similar villages scattered over the face of

England. It is perhaps a little richer and better run than most and it has for this reason been endowed with a poorer neighbour, Preston Stautney, which is decidedly worse off than the average village of the county. Together these two villages present a reasonably accurate picture of a rural community of the open or 'champion' country in the south of England around the middle of the fourteenth century.

Blakwater, then, was a medium sized village of some thirty families and a total population of about a hundred and fifty. Four of these families belonged to freemen paying rent to the lord of the manor but owing him no feudal service, other villagers were still all bound to the lord and had to do various works on his land in exchange for their cottages and strips of field. It did not seem likely that this would change rapidly since the landlord was William Edendon, Bishop of Winchester, and the Bishop, like most of his colleagues, was decidedly conservative in his attitude towards his tenants. He accepted that the commutation of labour services for money had already gone a long way in the English countryside and that—a point which caused him some distress—it was even to be found on his own estates. But he deplored the process, for social more than economic reasons, and it was well-known that his villeins would be unusually privileged if they were ever allowed to change their status.

The village lay about eight miles south-west of the King's road between London and Winchester; a broad river of mud in winter and of choking dust in summer. The traffic along this road was as heavy as on any in England, not of horsemen and pedestrians only but also of horse-drawn carriages, some of them, belonging to the families of the great magnates, vast and sumptuously decorated. Needless to say, no such carriage, nor indeed any sort of wheeled vehicle, ever found its way down the meandering footpath which led from the highway to Blakwater. Perhaps more surprisingly, very few of the villagers ventured any distance in the other direction. Not a single inhabitant of Blakwater had been as far as London, let alone to any foreign country. Only half a dozen had reached Winchester; the parson, the steward, the reeve and one or two of the more adventurous villagers. For the rest of the people, it was an expedition to walk even as far as the rickety wooden bridge which spanned the stream of Blakwater barely half a mile from the edge of the village.

They felt no sense of deprivation. The village was closely knit, introverted and, by and large, content with its condition. Certainly it conducted some minor trade with the outside world, exported a little wheat and cattle to the market, imported some cloth and the odd manufactured article. But such trade was conducted by foreigners from Winchester through the intermediacy of the reeve or steward; so far as the other villagers were concerned it seemed to have little or no relevance to their daily life. What happened in the next village, let alone the next county, was a matter, if not of complete indifference, at least of minor and academic importance. It was entertaining to listen to the tale of travellers in the same spirit as, to-day, one might crowd to hear the words of an astronaut; but only the romantic or the reckless actually want to go to the moon and the inhabitant of Blakwater was no more likely to want to go to London or to Calais.

Beyond Blakwater the track became even worse, winding circuitously over the hill to the little village of Preston Stautney some four miles away. It was inevitable that there should be a certain amount of intercourse between the two communities. The young people met in the woods to play games or to poach the lord's deer, sometimes they carried their games a little further and two or three of the families were linked by marriage. But on the whole the two villages kept themselves to themselves. There was no bad blood between them but the Blakwater folk tended to think themselves considerably better than their neighbours. For Preston Stautney was poor and small. Its land was nowhere near as fertile but this alone was not enough to explain the contrast. Twenty years before, indeed, it had been by no means so marked. But Sir Peter Stautney, the lord of the manor, was something of a wastrel. He liked to spend his time as a soldier on the Continent when he should have been tending his estates at home. Nor was his performance as a soldier likely to win much glory for himself or vicarious satisfaction for his tenants. Once, indeed, he had got himself captured by the French and his steward had had to sell a couple of villages and extract the last possible penny from the others before he could raise the necessary ransom. Discouraged by the lord's indifference the bailiff had grown slack and was believed to be lining his own pockets at Sir Peter's expense. The villagers took advantage of his idleness but were none the less resentful of what they felt to be his unfair exactions.

Preston Stautney, in short, was an unhappy village. It had dwindled to some fifteen families, a little over sixty inhabitants in all, and several of those who remained were talking of trying their luck elsewhere. It was, of course, against the law for a villein thus to desert his master but the bailiff would be unlikely to take any very vigorous steps to recapture the fugitive—especially if he had been softened up with a shilling or two in advance—and by the time Sir Peter discovered what had happened the refugee would be far away and beyond discovery. Not that they needed to go very far to be lost to Sir Peter. One of the free tenants at Blakwater was known to be an escaped villein from over the hill. He was a good worker and an honest man and the reeve had no intention of handing him back. Even if Sir Peter found out and complained to the Bishop he would not be likely to get much satisfaction.

For William of Edendon was one of the great magnates of the kingdom, attaching little importance to the protests of a country knight. More to the point, he was a capable and conscientious landlord, always ready to invest some part of his great riches in improving his estates and, though determined to get his due, never harsh or unreasonable in his exactions. He knew Sir Peter as an inefficient absentee, of interest only in that the chaos of his finances might make it possible to snap up one or two of his manors cheaply at some future date. The Bishop had put in the present steward some three years before, paid him well—fifty shillings a year, clothing, stabling for his horse, the use of part of the manor house and a peck of oats each day for his horse—and expected good service in return. The steward was responsible for seven manors in all but Blakwater was one of the largest and the most central and it was there that he had made his home. He came from somewhere the other side of Winchester, for the Bishop believed in putting foreigners in positions of authority on his manors, but he had been accepted by the villagers, if not as one of them, then at least as the next best thing.

On the whole the people of Blakwater thought themselves fortunate. But though they knew that they were more prosperous and more secure than their neighbours in Preston Stautney, from time to time they hankered wistfully after the greater freedom which the instability of the smaller village had incidentally bestowed on its inhabitants. For not only could the men of Preston Stautney leave their homes with impunity if they wanted to but the bailiff was always so short of ready money that it was easy, in exchange for a small payment, to get out of almost all the services which they were supposed to perform on the lord's demesne. Indeed, most of the former villeins had by now commuted all their services for life and worked on what little was left of the demesne only for a money payment. But though their neighbours might boast about their liberty, the Blakwater men were satisfied for most of the time that their own full stomachs and well-built houses made their lot the happier. Only now and then, when their reeve seemed more than usually exigent, did they wonder whether freedom might not after all be worth the price of poverty.

But it was not only in its steward that Blakwater was fortunate. The vicar, though not a particularly strong or dynamic character, was a good man; genuinely fond of his flock and conceiving it his duty and his pleasure to serve them diligently. It could have been of him that Chaucer wrote:

'A good man was ther of religioun,
And was a poure PERSON of a toun . . .
. . . He sette nat his benefice to hyre
And leet his sheep encombred in the myre
And ran to London, unto Seint Poules,
To seken hym a chaunterie for soules;
Or with a bretherhed to been withholde;
But dwelleth at hoom and kepte wel his folde.'*

The reeve, too, was fair and honest. He looked after the day to day administration of the village and understudied for the steward during the latter's frequent absences. He was one of the villagers, the brother of the thatcher indeed, and had been reeve for more

* 'A holy-minded man of good renown
  There was, and poor, the Parson to the town . . .
  . . . He did not set his benefice to hire
  And leave his sheep encumbered in the mire
  Or run to London to earn easy bread
  By singing masses for the wealthy dead,
  Or find some Brotherhood and get enrolled.
  He stayed at home and watched over his fold.'

than twenty years. Now he was an old man and he had told the steward that he wanted to retire at the end of the year. In theory his successor would be elected by all the tenants of the village at a Manor Court but in practice the steward and vicar between them made sure that their candidate was the only one to be nominated. The identity of the new reeve was already decided on and known to all the village. It was to be Roger Tyler; descendant of tilers perhaps, but with no knowledge of the trade himself. Instead he was said to be the best handler of cattle in the neighbourhood and a sensible, determined man whose authority would willingly be accepted by the other villagers.

As befitted one of the richest of the villeins, Roger Tyler lived in a large, three-bayed house with matting on one of the floors and, a feature of rare luxury, a strip of oiled linen-cloth over one of the four windows. With him lived his old and invalid father, his wife, his sister and his four children—three sons, the eldest aged fourteen, and a girl of six. The family lived well, eating meat more often than any other household in the village except that of the steward. Certainly Roger's standard of living was higher than the parson's. Eggs were to be had most days, fish at least once a week and cabbages, leeks, onions, peas and beans were all available in season. For the main meal of the day it would be quite usual to eat a vegetable gruel, rye bread, meat and a piece of cheese, washed down with cider or a thin beer made without hops. He had a few fruit trees as well: apples, pears and a medlar, and he took a share of the walnuts and chestnuts from the garden of the manor. In winter, of course, things were harder, but there was almost always a piece of salted bacon in the house. Unfortunately salt was so expensive that even Roger Tyler was forced to skimp and the bacon was often rancid and almost uneatable long before spring arrived.

Things were different next door where Roger's widowed aunt lived alone. Roger had tried to persuade her to join his family but she valued her independence too high. In Chaucer's words again she was:

‘ A poure wydwe somdel stape in age
Was whilom dwelling in a narwe cotage
Beside a grove, stondynge in a dale . . .
. . . Thre large sowes hadde she, and namo,

Thre keen, and eek a sheep that highte Malle,
Ful sooty was hir bout and eek hire halle,
In which she eet ful many a sklendre meel . . .
No wyn ne drank she, neither whit ne reed;
Hir bord was served moost with whit and blak—
Milk and broun bread, in which she found no lak—
Seynd bacon, and somtyme an ey or tweye;
For she was, as it were, a maner deye.'*

Where Roger's family slept on bags of flock, she made do with
a few handfuls of straw on the mud floor; cider and beer were an
unknown luxury in her house and, as against Roger's well-
organised messuage and commodious barn where he stored
fodder for his cattle, she had only a tumbledown shed where her
pigs jostled for standing room. But she never complained about
her lot and comforted herself with the thought of her good luck
compared to those unfortunates at Preston Stautney who often
had not got a single pig or even a chicken to their name. Besides,
her relationship to Roger gave her a standing among the élite of
the village: a select group which included the families of such
worthies as the manorial clerk, the miller and the reeve.

Though Roger himself made a point of keeping the domestic
animals out of the house this was by no means an invariable rule.
In some of the houses goats, sheep and sometimes even cows
lived jumbled up with the family, spreading their fleas amid the
soiled straw and adding their smells to the rich compound which
the medieval household could generate even without such extra
help. Washing was a luxury and probably weakening to the constitution—to be indulged in with caution and only at long intervals. Bathing was unheard of. Needless to say, in such conditions,
almost everyone had some sort of skin disease. Eye infections

* '. . . . . . a poor old widow
    In a small cottage by a little meadow . . .
    Three hefty sows, no more, were all her showing,
    Three cows as well; there was a sheep called Molly,
    Sooty her hall, her kitchen melancholy,
    And there she ate full many a slender meal . . .
    She drank no wine, nor white nor red had got,
    Her board was mostly served with white and black;
    Milk and brown bread; in which she found no lack;
    Broiled bacon or an egg or two were common,
    She was, in fact, a sort of dairy woman . . .'

were also common and the lack of green vegetables led to a certain amount of scurvy. But in spite of the risks which the lack of hygiene involved for the new born baby or the nursing mother, the average villager was still reasonably healthy: his complaints more irritating than dangerous. The older inhabitants liked sometimes to recount tales which they had heard from their fathers about fearful pestilences which carried away great numbers of the villagers but the young were openly bored by this tedious romanticising.

To the casual visitor from the present days the first impression of Blakwater might well have been that of a little village of some green upland in Swaziland or Zululand. The stone church with its round Saxon tower and Norman nave would have struck an unfamiliar note but the mud and wattle cottages with roof of reeds or hide and smoke seeping from every pore were superficially very like those to be found to-day in many of the less developed countries. The manor itself, with its large timber hall, where the court was held, its thatched wall of earth, and the big room above the gate reserved for the visits of the lord or his representative, was by far the most conspicuous group of buildings in the village. Within the wall it had a dovecot, a large fish pond and a well-stocked orchard, thatched hay-ricks, barns, stables and hen-houses: all the appurtenances, in fact, of a well run farm. The water mill lay just outside the walls; stoutly built on a frame of timber and sheltering the brand-new mill-stone from Northern France which was the miller's pride. All the land around was part of the lord's demesne.

The church was the other side of the manor with the parson's house beside it. Then came a group of houses belonging to the richer villeins, Roger's prominent among them, another row of houses similar in size but with rather less in the way of garden and out-buildings, where the less important villeins lived and finally the one-room huts of the cottars. For the most part the freeholders also had their houses in this part of the village. These paid rent to the lord instead of doing work for him and felt themselves to be far superior to their fellow villagers who were still bound to work on the demesne an average of three days a week.

But as Blakwater was richer than Preston Stautney, so the villeins of Blakwater were richer than the free tenants with the solitary exception of the miller who somehow contrived to unite

independence with affluence. Their poverty was a source of constant chagrin to the freeholders, but, since the Bishop clearly intended that they should do no better while he remained their lord,
they saw little hope of remedying the position.

Finally, on the fringe of the village, a ramshackle hovel provided shelter of a sort for poor Mad Meg; deformed from birth,
shunned by her contemporaries and now grown crazed in squalid
loneliness. Some said that she was a witch and the children used
to enjoy chanting rude slogans outside her hut but nobody
seriously believed that she could make successful mischief.

In spite of its position nearly two hours' walk from the highway,
Blakwater was by no means cut off from the outside world. Down
the little track came every kind of pedlar and huckster, free
labourers looking for a new home, quack doctors, pardoners and
friars, travelling shoemakers, the occasional minstrel, and seamen
or voyagers taking a short cut across country to or from the
Hampshire coast. It was one of these last who told the villagers
that a dreadful plague was raging on the continent of Europe. He
had not actually seen anything of it himself, nor indeed met anyone who had, but in Bordeaux, which he had lately visited, the
port had been buzzing with horrific stories. The villagers were not
particularly impressed. Where was this plague then? In Italy. And
where was Italy? Rome they had heard of but the sailor did not
know whether it was in Rome or not. 'Poor folk,' they muttered
perfunctorily, and let the matter slip from their minds.

A few weeks later—it must have been in March or April, 1348
—they heard the same story again. This time it came from one of
the serfs at Preston Stautney, a man who had accompanied his
master to the war and was now on his way home. He too had not
seen anything himself but he claimed actually to have spoken to
a Franciscan friar who had been in Avignon a few weeks after the
plague arrived. He told of whole families wiped out, of pits filled
with dead, of black clouds of lethal smoke destroying all who
smelled or even saw it, of men erect and healthy at one moment
and dying in agony at the next. Again the villagers nodded their
heads sadly. England might not be paradise but at least it was a
safer and better place than those unknown and dangerous
countries across the sea.

It was not till the beginning of September that a report came that the plague had crossed the Channel and there were now victims on English soil. The news still made surprisingly little impression on the villagers. A plague in Dorset seemed to have little more relevance to their lives than in France or Italy. The harvest was in full swing and the only thing that really mattered was that they should get in the lord's wheat in time to deal with their own before it rained. Any other consideration would have to wait. And when the harvest was safely in and they had time to concentrate on the news they still saw little to discomfit them. The plague was now said to be moving away towards the west. It was ridiculous to suppose that the Bishop would let it get any closer to his diocese.

Then came that Sunday in October when the parson mounted even more slowly than usual to his pulpit and, in a voice that he seemed to control with difficulty, told the congregation that he had received a letter in Latin from the Bishop and was now going to read a translation of it. 'A voice in Rama has been heard . . .' began the message, and went on to tell of the horrors which the plague had inflicted on people all over the world. The villagers shuffled their feet and looked furtively at one another. What were they supposed to do about it? '. . . this cruel plague,' the parson read on, 'as we have heard, has already begun to afflict the various coasts of the realm of England. We are struck by the greatest fear lest, which God forbid, the fell disease ravage any part of our city and diocese . . .'

Roger Tyler drew in his breath and gave a look of shocked dismay at his wife and children beside him. Could it be then that the threat was real; that this dreadful thing could happen here in Blakwater? The parson droned on about psalms and penances but Roger barely listened. His mind was filled with a sense of dawning horror, a fear of something unknown and awful yet, in a curious way, painfully familiar. Perhaps, without realising it, he had for several months been preparing himself for the news which had now arrived. He felt suddenly cold and pulled his tunic closer around him with a half-unconscious gesture.

It might have been the fact that the parson was speaking in church so that no one could argue or ask questions which made the whole thing seem more fearful. Somehow it was not quite as bad when he talked it over with his friends in the churchyard after the service. The Bishop didn't say that the plague *was* coming, merely that it might. Evidently he believed that the saying of penances might avert the danger—well, the villagers could say plenty of those. After all, a plague was something which happened in big cities, not peaceful country villages. Probably, too, there was a lot of nonsense being talked about its severity. Everyone knew how exaggerated reports of this kind could be.

So the villagers comforted one another and, on the whole, they did it well. There was unusual activity around the church for a week or two but after a time even this began to flag and things went on much as before. The Bishop sent the parson another letter but he had nothing new to say. He wanted them all to pray, of course. Well, that was what Bishops were for and they would do as he said, but they were not going to panic because a lot of foreigners were dying at the other side of the country. They questioned avidly any new arrival at the village but the plague never reached any place of which they had heard.

Then, a week before Christmas, a pedlar stumbled into the village with the eager gloom of one who has bad news to break and means to make the most of it. He had been travelling to Winchester, he said, and had just reached the point where the track to Blakwater met the highway when he saw a group of men and women riding furiously in the other direction. Two or three of them were people of distinction, the rest servants. One of the servants stopped near him to adjust the baggage on his horse and the pedlar asked him what the hurry was. They were fleeing from the plague, was the answer. Hundreds were dying every day in Winchester, the graveyards were overflowing, there was fighting in the streets. He galloped off after his master, leaving the pedlar to look after himself as best he could.

There was no easy comfort now for the villagers. If the plague was in Winchester to-day then it might be in Blakwater to-morrow. The parson organised all the men of the village into a barefoot procession, carrying crosses and singing psalms. They marched around the village and ended up in the churchyard. But, as the parson privately admitted to Roger, he did it more to keep the people busy than in the hope that it would do much good. If God had decided that his people must be punished he was not likely to be deflected from his purpose at this late hour. 'If it must come, it must come,' he concluded gloomily, looking down the track that led to Winchester. Roger looked too; but what he was looking for he did not know: a sick man, perhaps—or the spectre of Death riding on a black horse?

The villagers were in an odd mood all that week. There was a strange, febrile gaiety. Everyone laughed and joked a lot and avoided talking about the dread that was uppermost in their minds. The village was always a friendly place but there was now an unusual sense of comradeship. People spontaneously helped each other; even Mad Meg was treated with unusual respect. The twelve days of Christmas were observed with all the usual jollification: more than usual, indeed, since everyone except the children was behaving as if he was acting a part and, like every amateur actor, was badly over-playing it. There was some suggestion that the steward should cancel the traditional dinner for the tenants on Christmas night; 'At Christmas we banket, the rich with the poor,' as Thomas Tusser was later to put it. But the outcry was immediate and the dinner was held with all its usual bawdy fun and the election of a lord of Misrule.

When the plague came in the end it was not by way of Winchester but through the back door from Preston Stautney. It had been a mild, wet winter and Bartholomew Thomasyn was barely recognisable for mud when he staggered up the track that linked the two villages. He had married a girl from Blakwater and so was well-known there but the friendly greetings died on the lips of the peasants as he stammered out his news. His wife was dead, he said and their little daughter with her. At least six other villagers were dead or dying. With nothing left to hold him to his house, he had fled. Soon others would be following his example. The village was doomed and all who stayed there would perish. He had not slept for three days nor eaten for twenty-four hours. His father-in-law led him away to get some rest and the group of peasants that had clustered around broke up with scarcely a word.

That afternoon the reeve walked quietly around the village asking a few of the wiser or senior inhabitants to come to a meeting in the hall of the manor. The steward was away on his travels but the parson was going to be there. Roger, of course, was among those invited. The group duly assembled but, when they were all seated around the table, no one seemed to have any idea what to say or do. The parson should have taken the lead but he looked ill and half-asleep and contributed nothing to the discussion while the reeve muttered some prepared preamble about the need to do something urgently and then looked helplessly around for contributions.

In the end Roger spoke up. Everyone who knew anything about the plague, he said, agreed that it was fatal to have any contact with the victims. The village, therefore, must cut itself off. Travellers should be forbidden to use the road through the village; if they wanted to by-pass it through the fields then they were welcome to do so but they should come no closer. Above all the village must have nothing to do with Preston Stautney. If any-one from there tried to enter Blakwater, they must be turned away—by force if necessary. This was not uncharitable for, after all, there was nothing that could be done to help. Anyway, charity began at home. Their first duty was to their wives and children.

No one spoke up against Roger's plan but one of the elder villeins stirred uneasily at the end of the table. What about Bartholomew Thomasyn, he asked; was he to be allowed to stay? 'He must go,' put in Roger quickly, before anyone else could speak. He knew as he spoke how harsh his words must sound to Bartholomew's father-in-law who was also at the meeting but he thought of his wife and children and knew that he was right. The reeve would back him and he could over-persuade the other villeins but he was nervous lest the parson should oppose him and argue that it was their Christian duty to help the sick. He glanced down towards the end of the table but the parson had his head sunk in his hands and gave no sign even of listening to the dis-cussion. Bartholomew's father-in-law protested but without much conviction. To all of them the peril seemed too great to leave room for sentimentality.

Suddenly the parson lurched to his feet. The peasants fell silent and looked expectantly towards him but, instead of speak-ing, he turned away and staggered through the door. Was he overcome by anger at their lack of charity, wondered Roger nervously? They watched him totter down the path towards the manor gate, reeling from one side to the other and seeming every instant about to fall. 'Parson had a bit too much to drink?' speculated one of the villagers. As he spoke the parson pitched forward on his face, tried to drag himself to his feet, then fell forward again and lay still. In a few seconds Roger was at his side. His breath was coming with a heavy wheezing noise, his cheeks were so hot that Roger snatched away his hand in alarm as he touched them. They carried him to his house and pulled the clothes from his twitching, fevered body. Under both arm pits and in his groin red boils were growing: still small but not so small that those who saw them could doubt that they were the dreaded plague buboes about which they had heard so much.

Without looking at each other, without a word, the villagers slipped from the parsonage and fled to their own houses. Against the immense peril of the plague they had no recourse save that of prayer. Yet now the sickness of the parson seemed to have cut them off even from that ultimate hope. If God first struck down His chosen servant how terrible must be His wrath against the others! For a few hours they skulked indoors, scarcely venturing even to look outside. Everywhere, it seemed, the poisoned breath of the plague must be awaiting them. It was almost dark before Roger pulled himself together and walked out into the silent lane. He went from house to house, calling to the inhabitants. No other case of the plague had yet occurred. At the parsonage the parson had dropped into a restless sleep. He tossed and turned but his fever had grown a little milder. Could it be that Blakwater would escape lightly, that God would content himself with this dire warning and now avert his wrath. There seemed at least a ray of hope.

This happened on a Monday. On Tuesday there was no new case and the villagers began to creep cautiously from their houses and to talk together in hushed voices. On Wednesday there was still no further outbreak. The parson's buboes had swollen and were now inflamed and painful but he himself had recovered consciousness and showed no signs of imminent decease. In sharp reaction to their earlier despair a wave almost of euphoria overcame the villagers. Surely the danger of a worse outbreak must fast be passing? Most of the peasants went off to work in the fields and, generally, life was returning to normal. Seeing Bartholomew Thomasyn outside his father-in-law's house Roger remembered his plan for sealing off the village. With the plague already inside, there was little point in such precautions.

Wednesday was cold but clear. Roger rose at his usual time, looked anxiously at his family and saw with relief that all were well. Another night safely passed: he walked outside into his garden. His aunt's house was quiet and no smoke came from the fire. Odd, she was usually up before him. In sudden apprehension he ran to the door. As he approached he heard a low moaning from within. The old lady was sprawled in a heap on the ground; she must have been overcome on her way to seek help. Her face was haggard, her eyes sunken and blood-shot. Her swollen tongue protruded from dry, cracked lips. She was barely conscious but aware that Roger was beside her. 'Water!' she croaked, in a whisper that hardly reached her nephew's ear, 'Water!' When a pot of water was brought, she drank it down greedily; she was unable to control the movements of her tongue and, in spite of Roger's efforts, a lot of the water dribbled down her front on to the floor. When the pot was empty she fell back exhausted, breathing stertorously but apparently a little the better for her drink.

Roger left her house to break the news to his wife. As he stepped from the hut he heard a harsh scream from behind him. The wife of one of the villeins burst out from her house into the road. In her arms she carried her little child; yesterday a healthy, cheerful boy of four months, now transformed in a few hours into a distorted and pain-racked caricature. 'My baby,' was all she could cry. Again and again: 'My baby!' 'My baby!' 'My baby!' Her husband ran after her and, with Roger's help, mother and child were hustled back into their house. Even as they got inside the door the child stiffened itself in a final spasm of agony and lay back dead.

Almost stunned with horror Roger went back into the road. Was there any chance that the parson might be better and able to give consolation to the still hysterical mother? He walked quickly to the parsonage. As he entered he staggered back, overcome by the horrifying stench. The parson's buboes had burst. His eyes wide open, his fists clenched, he was lying dead, staring blankly upwards from the pool of suppurating black filth which had oozed from the open boils. Roger turned and fled. Once in the garden he knelt and was violently sick.

With the parson's death and the steward away the steward's clerk was the only villager left who knew how to read and write. Usually he went with the steward on his travels but this time he had luckily stayed behind. In the name of the reeve he now wrote to the Bishop explaining the disasters which had overtaken the village and pleading that a new priest might be sent them as soon as possible. The letter was taken to the highway by one of the villeins and entrusted to the first respectable-looking traveller who came along. Now the village could only wait and hope.

The next two months were an almost uninterrupted nightmare. Sometimes two or three days at a time would pass without any new victims and the hopes of the villagers would begin to rise, but always in the end the disease struck again. One by one they sickened and died: the survivors kept the tally of the dead and wondered secretly who would be the next to go. It seemed that the hunger of the plague would only be satisfied when the last inhabitant had followed his parson to the grave. The old reeve was one of the first to die, leaving the village with no sort of leadership. All the men who had the courage and the strength rallied to the hall of the manor and elected Roger their new reeve. The court was not properly constituted in the absence of the steward but, in the circumstances, no one was disposed to worry about formalities. Apart from this one burst of corporate activity, the village lapsed into total apathy. Nobody tended the fields— for who would be left alive to reap the harvest? The cattle were neglected; the flimsy houses began to fall into disrepair; men and women lost all interest in their own appearance and lurked fearfully in their houses as if afraid to face the open air.

Almost the only occupation which aroused any interest was burying the dead. In the parson's absence, the steward's clerk used to read the service. After ten days or so of this, however, an itinerant friar turned up on the way to his religious house at Romsey. He promised the villagers that he would stay at Blakwater until a new priest arrived or the plague was over and he was as good as his word; tending the sick and laying out the dead with a fearlessness which quickly won him the respect and affection of the people. Then a new problem arose. In the first three weeks more than twenty people died and the old churchyard, crowded even before the epidemic, was quickly choked with new graves. Even if it had not been, Roger had a theory that it was dangerous to bury the plague victims so close to the centre of the village. He asked the friar to consecrate a new plot of land a few hundred yards away on the edge of the lord's demesne. At first the friar refused; the plague could not frighten him but the anger of the Bishop if a new churchyard was opened without his permission and the payment of the usual fee was quite a different matter. However Roger promised that the fee would be payed and everything regularised in due course and, in the end, the friar grudgingly agreed.

The very next day a chaplain of the bishop's rode into the village. The letter had been received and would be acted on, he told Roger, but there was no hope that the new priest would be in Blakwater for another three weeks at least. At the moment the Bishop had more than eighty benefices vacant in the diocese and, though he was doing all he could to fill them, some delay was inevitable. The chaplain looked coldly at the friar and still more coldly at the new graveyard but, since he could do nothing to remedy the matter, wisely held his peace. The ugly sullenness of the villagers probably warned him that it would be unwise to push them far.

When the chaplain rode on to Preston Stautney, Roger went with him so as to see how his neighbours were faring. Only then did he realise that, however badly Blakwater had suffered, others had fared still worse. The community had disintegrated. Of the sixteen or seventeen houses only four seemed still to be inhabited. The door of the church was broken down and somebody had been chopping up the stalls, presumably for firewood. Of the parson there was no trace at all, unless a large mound of freshly-dug earth in a corner of the churchyard covered both priest and flock. The only people they could find were a couple of old women sunning themselves forlornly outside their houses. All the others were dead, they said, dead or run away. The chaplain cross-examined them in an effort to get some rather more precise information and in the end established that at least a dozen villagers had taken to the woods in the hope of escaping the plague. But whether they were still alive or had been struck down in their flight, the old hags neither knew nor cared.

Soberly Roger returned to his home. He had seen so much suffering in the last few weeks, had felt so much pity and so much fear, that it seemed he had no emotion left which could be squeezed out for the sake of these further victims. Indeed, as he walked down the hillside to Blakwater he caught himself in a mood of self-congratulation at his own light escape. Uneasily he crossed himself and dismissed the dangerous thought from his mind. He had cause to remember his gesture and the moment of disquiet which had inspired it. When he arrived home he found his eldest son groaning with pain, vomiting almost continuously and in a high fever. The boy died after four days of intolerable suffering.

Even before he was in his grave Roger's only daughter and his wife were on their sick beds. The former was one of the very few who were infected by the plague but still survived—her life was in great danger for several days but by some freak of chance the buboes proved less malignant than in other cases and subsided or suppurated harmlessly. Roger's wife fought for her life for more than a week, clinging on tenaciously even when her body had been reduced to a shattered and malodorous hulk. In the end she succumbed and Roger cursed the god who could bring such misery on his defenceless servants.

Impotent, resentful, panic-stricken: the villagers were in a mood to revenge themselves on any target which came within their range. Poor Mad Meg provided an easy victim. Someone had met her by night conversing suspiciously with her obviously diabolic cat. Someone else had seen her lurking near the well—armed with poison without a doubt. A crowd of villagers worked themselves into a drunken frenzy on beer looted from the house of the ale-brewer and marched indignantly towards her house. Mad Meg heard them coming and slipped away into the woods. Probably she would have escaped their clumsy pursuit if one of the peasants had not seized hold of her cat and, brandishing it by the tail, smashed its head against a rock. In hysterical defence of the only living creature that had shown her any trace of love, Meg ran out from her hiding place among the trees. The villagers attacked her with sticks and stones and battered her to death in the clearing outside her miserable hovel.

Even the longest nightmare must end. By the time that the new parson arrived in early March the worst was over. The plague lingered for another two months but its full ferocity was past. A gap of four days occurred before the next case, then of five, then of a week; by the beginning of August there had been no new attack for nearly two months and the villagers could feel themselves safe. Thirty-eight of them had died, three others had been infected but had recovered, poor Mad Meg also had her claim to be a victim of the plague. Little by little the survivors began to look about them, to realise that they were still alive and likely to remain so, to pick up the pieces of their lives again.

They had plenty to do to keep themselves occupied. All the work in the fields had been neglected for more than six months and now, with a greatly weakened labour force, they had to make good the wasted time. But there were compensations. The same amount of land and cattle was now available to be shared out among fewer people; this meant that the work was harder but also that the reward was greater. Roger, who had considered himself one of the most over-worked villeins on the manor, was pressed by the steward to take on half his neighbour's land at a nominal rent. Anything, the steward pleaded, was better than that it should go to waste. Reluctantly Roger agreed and found to his surprise that, with some hired help from one of the freemen of Blakwater, he could manage the extra land quite easily. Two or three other villeins also took on extra land and found themselves increasingly prosperous as a result.

However much new land the more energetic villeins had taken on, it would not by itself have been enough to fill the gaps left by the plague. But Preston Stautney's loss proved to be Blakwater's gain. The tenants of Sir Peter, who had escaped death by taking refuge in the woods, now saw little to attract them in their stricken village with its barren land and thriftless landlord. Some fled to more distant parts to make a new life but a few—four men in all with what was left of their families—arrived one day at Blakwater and appealed to the reeve and steward to let them settle. They said they were even prepared to give up their status of free men and to bind themselves to tender service to the Lord Bishop in exchange for a house and land.

Roger was anxious to take them in but the steward was less certain. The King had not yet passed his new laws forbidding the movement of free labour but it was, to say the least, unneighbourly to attract away peasants from a nearby village. Besides, though the four men claimed to be free, the steward had some private doubts whether they could prove their status in a court of law. But labour was short and expensive and the harvest had to be got in. In the end the steward agreed that they could stay until the Bishop's representative paid his next visit and that the question would then be put to him. By the time the latter did come the men were firmly installed and it seemed a pity to disturb an arrangement which was working so satisfactorily. It was decided that they could stay, at least unless Sir Peter protested strongly. Since nobody thought fit to tell Sir Peter where his errant tenants were to be found, such a protest was never made.

By means such as this, the bailiff was remarkably successful in assuring an adequate supply of cheap labour and the Bishop of Winchester lost little financially. In the year before the plague the Bishop had gained an income of some £70 from the manor of Blakwater: £20 for the profit from farming the demesne and £50 in rents, fines and various court perquisites. In the year of the plague the pattern altered greatly. The profit from the demesne almost vanished and rents dwindled dramatically, either because the tenants were too poor to pay them or because nobody was left alive. But this was more than offset by a sharp increase in the income from fines payable on the estates of the dead or from property which escheated to the lord because no heir could be found. The Bishop ended the year with a profit on his manor slightly greater than the year before. This economic ebullience proved illusory. In the next twelve months higher labour costs, and uncertain markets continued to depress the profit on the demesne. Rents recovered, but by no means to the level which had prevailed before the plague. In particular the water mill, usually one of the most profitable items in the bailiff's accounts, stood empty until the middle of 1350. The windfall which had come from fines on the estates of the dead could not be repeated a second year. The Bishop still continued to break even, but only just. It was another three years before the income of the manor returned to its former level.

The stoutest opposition to accepting the peasants from Preston Stautney came from the new parson. To him anything which differed even slightly from the past was to be distrusted if not deplored and mobility of labour was obviously contrary to all the established principles of good government. But Roger, now officially reeve, cared little for the parson's objections. What had he done for them when things had been at their worst? If the peasants owed a debt to anyone it was to the travelling friar who had discreetly vanished when the new parson rode into the village from Winchester. The villagers listened sullenly when the parson denounced the friar's presumption—why, after all, should they care for a church which had so conspicuously failed to protect its flock. They had by no means lost their faith in God but their enthusiasm for God's ministers on earth had worn thin. When one of those wandering gangs of brigands which seemed so omnipresent in the years that followed the plague broke into the church and stole the silver cross, everyone was profoundly shocked. But when the same gang or another one stole the parson's pig the villagers laughed heartily and wished them luck.

By the end of 1350, to the casual visitor, Blakwater must have seemed almost back to normal. There were some new faces of course, an unusual number of widows and widowers, empty places in church. Sad little pilgrimages to the new graveyard outside the village had become a part of the daily routine. But only one house and, of course, Mad Meg's shack remained untenanted and the latter had almost vanished under the assaults of wind, rain and mischievous children. The fields looked much the same as ever and the water mill was grinding away merrily. The dovecot of the manor had been repaired and the fish pond restocked. But any less cursory study would quickly have revealed that the village was like a man whose gangrenous arm had recently been cut off. In strictly physical terms the wound was more or less healed, but a few months could not eliminate the shock or sense of deprivation. There were still starts of pain in the vanished limb and the victim walked in dread that the gangrene would re-emerge and his sufferings start all over again.

One day when the harvest was over Roger walked over the hill to Preston Stautney. The grass grew thick in what had used to be the main street, the wall around the manor had collapsed, the mill was derelict. As he pushed his way among the houses he saw that a few of the houses were inhabited, a corner of the fields was still being cultivated and someone had made a pathetic effort to clear the churchyard of the worst of its weeds and brambles. But in most of the houses the roofs had fallen in and the walls were beginning to tilt at crazy angles. He made his way to the church. The door had fallen from its hinges: birds were playing in the roof; a strong, pungent smell suggested that a fox had taken up residence beneath the wreckage of the pulpit. A pig was snuffling and rooting among the graves. With a shudder of disgust Roger drove it away; then turned and left the village without a backward glance.

He was not a happy man. He had lost a son and his beloved wife. He had seen horrors that would linger with him all his life. But he still had three children left; he was luckier than some.

Hard work and the knowledge that he had an important role to play had helped him over the last months. Blakwater was at least a living village, Preston Stautney was a village of the dying, if not already of the dead. He turned his face towards the living with sadness, with fear but also with a kind of gratitude. The nightmare was over. The pain remained but there was, after all, a great deal to be said for being alive.

FOURTEEN

The Toll in Lives

In Blakwater, thirty-eight people died out of a total of about a hundred and fifty; close to a quarter of the population. In Preston Stautney things must have been worse; probably nearly half the villagers succumbed. Which of these villages was nearer the national average? Can, indeed, any national average be established? Did a higher proportion of the population die in England than, for instance, in France or Italy? And how large was the actual death roll? Did a million English die? Two million? Three?

To none of these questions is a categoric answer possible but, now that the geographical tour of Europe has been completed, it is at least possible to hazard a few guesses. The most ample material on which an estimate can be based is certainly to be found in England but even here the base is shaky and deductions hazardous. It is possible to arrive at a wide variety of conclusions by differing but reasonably valid lines of argument, and exceedingly hard to establish which, if any, is the best one.

The first and, in some ways, most perplexing problem is the size of the total population in the middle of the fourteenth century. The main difficulty is that no attempt at anything approaching a general census was made between Domesday year and the poll-tax returns of 1377. Nor did even these attempt to cover all the counties of England or all kinds of men. Nevertheless it is possible to hazard a reasonably confident guess that the population of England in 1086 was something near 1.25 million,[1] and that, by 1377, this had risen to about 2.5 million. If it were permissible to assume a steady increase of population between these two points then it would, of course, be easy to arrive at the approximate size of the population at any given date. But this is

very far from being the case. On the contrary it is now established
with a fair degree of certainty that the population rose to a peak
about 1300 and then stagnated or even declined in the first half
of the fourteenth century.

Exactly what caused the economic decline between 1300 and
1348 and how far, if at all, it was reflected in a reduction of the
population has been the subject of much debate. Dr Titow has
cited evidence from the Winchester Account Rolls to show that
the great famines of 1315 to 1317 were the turning point.[2]
Though, in some areas, the recession seems to have begun ten
years or more before, in general the statement seems valid. The
famines themselves cost many lives but, in the palmier days of the
thirteenth century, this loss would quickly have been made good.
In the fourteenth century no such recovery took place. In a
highly important article Professor Postan has demonstrated that,
while wages rose gradually and taxation did not decline, there
was a fall in agricultural output and in exports.[3] The explanation
must be a smaller force of labourers to share the pay packet. There
is evidence to the same effect to be found in the narrowing wage-
differential between skilled and unskilled labour and the withdrawal from previously cultivated land. 'The contemporaries
obviously believed that they were living in an age of contracting
settlement,' commented Postan, 'and there is no reason why we
should not accept their belief at its face value.'

The population in 1348 was, therefore, certainly little greater
and probably less than it had been in 1310. But this does not tell
one how large it was. Seebohm was the first historian to grapple
with the problem in anything approaching modern terms.[4] He
visualised 1348 as a peak, attributed the rapid rise in the preceding
century largely to the immigration of fishermen and manufacturers of woollen cloth, and concluded that the population just
before the Black Death was in the region of five million. Thorold
Rogers promptly countered with the contention that England
could not possibly have supported a population of five million.[5]
He analysed the farm accounts of eight thousand bailiffs and deduced from the production figures that the population of England
and Wales together must have been somewhere between two and
two and a half million. After a delay for cogitation Seebohm
replied challenging Rogers's figures for corn production.[6] The
wrangle was there allowed to rest. For the next seventy-five years

population estimates varied between these two points, usually inclining towards the higher.

In 1948, Professor Russell for the first time brought highly sophisticated statistical techniques to bear on the problem. His conclusion was that the English population in 1348 was some 40% larger than at the time of the poll-tax, in round figures about 3.7 million.[7] His graphs and tables are awe-inspiring but behind his arcane statistical manœuvres the validity of his conclusion rests to a great extent on the gratifyingly comprehensible assumption that the average medieval household contained only 3.5 members and not five as had previously been assumed. The significance of this figure lay in the ratio which it established between land-holders, whose deaths were recorded, and the rest who usually died unchronicled. If his index figure were to be raised by even half a person per household, the total population would certainly be increased to well over four million. An accurate index figure must therefore be fundamental to any calculation.

Professor Russell justified his somewhat dramatic departure from accepted theory by evidence drawn from inquests of enclosures, poll tax lists and other sources.[8] This is far from being unchallenged. The counter-argument, in its simplest form was that the Russell household unit contained only the nucleus of parents and children. But there is good reason to include other members, such as a retired father, unmarried brothers or sisters, servants and sometimes even sub-tenants.[9] Roger Tyler's household included seven people in addition to the tenant himself—a large unit, certainly, but by no means improbably so. Professor Russell's calculations, it is claimed, were based on an extremely limited number of cases and his evidence drawn largely from the period which followed the plague. If the 3.5 index were applied to the figures established for 1311 then, Dr Titow has pointed out, it would 'postulate a society in which male persons over twelve years of age constituted 59% of the total population.'[10] Undoubtedly there is considerable variation between one period and another but Dr Krause, who analysed Professor Russell's calculations with thoughtful distaste, cannot accept that in the fourteenth century the index fluctuated more than between 4.3 in a period of low childbirth and 5.2 when the rate was high.[11]

Faced with statistical juggling of this kind the layman is apt to feel a sense of baffled helplessness, leading often to blind acceptance of the latest theory which happens to have been propounded. He would do well to remember Professor Elton's expression of lapidary wisdom:

'Those determined to put their faith in "sophisticated" mathematical methods and to apply "general laws" to the pitifully meagre and very uncertain detail that historical evidence often provides for the answering of just such interesting and important questions, are either to be pitied because they will be sinking in quicksand while believing themselves to be standing on solid earth, or to be combated because they darken counsel with their errors.'[12]

Professor Russell is far from having accepted these strictures on his theory.[13] But he is too serious a scholar to maintain categorically that his must be the correct solution. The question remains open. In so far as any consensus can be said to have evolved it would probably be that the total population could have been anywhere within a range of which Russell's 3.7 million would be the lower point and 4.6 million or so the higher. A total of 4.2 million has no more precise justification than any other but is certainly no less plausible and is a convenient central point from which to work.

Of this 4.2 million, how many died?'Only one in ten survived,' says one chronicler; 'three quarters perished,' says another; 'four fifths,' a third. Few estimates fall as low or lower than a half. Such lurid speculations, of course, contain little of interest to the statistician; enough cases have been established where the estimates of the chroniclers were palpably impossible to dispel any lingering belief that the man on the spot knew best. But to arrive at a more rational figure is not easy.

One much favoured method is to seek to calculate from the ecclesiastical records the number of the beneficed clergy who died, to establish this as a proportion of the total and then to apply the same or some related percentage to the lay population. It was the application of this technique which led Cardinal Gasquet to claim that fully 50% of the population perished in the two years beginning in July, 1348.[14] The imperfections of the method have already been discussed.[15] Cardinal Gasquet himself was led badly astray by such deficiencies and even the more evolved workings of Professor Hamilton Thompson and Dr Lunn leave certain pockets of uncertainty.

Nevertheless studies of this kind can produce interesting and highly relevant results. Hamilton Thompson[16] and Lunn[17] between them have established the mortality rate of beneficed clergy in ten of England's dioceses. The figures are remarkably consistent, ranging between just under 39% for York and 39.6% for Lichfield to 47.6% for Bath and Wells, 48.5% for Ely and 48.8% for Exeter, Winchester and Norwich. On this basis it is reasonable to assume that something close to 45% of all parish priests died during the plague. Similar statistics based on twelve of the more important monasteries show a surprisingly similar rate among the monks, 44% of whom perished.

But though these figures are undoubtedly relevant to the problem of the total casualties caused by the Black Death, exactly how they should be used is harder to establish. It is as certain as any medieval statistics can be that, for England as a whole, the mortality rate among the people was lower than 45%. Applying the ratio between dead clergy and dead people referred to above[18] one must conclude, on the other hand, that it cannot have been lower than 34% or so—say a third to avoid any false impression of exactitude.

Professor Russell, who found the figures for clerical mortality difficult to reconcile with his own very low estimate for deaths among the whole population, tried to overcome his difficulty not, as might have been expected, by assuming the existence of a larger differential between the two categories but by suggesting that the former figures were incorrect. 'With some reluctance,'[19] he reached the conclusion that Professor Hamilton Thompson, that 'careful scholar who knows ecclesiastical practice so well,' had nevertheless been guilty of some fairly elementary blunders. But since the blunders whose presence he suspected were specifically those which Hamilton Thompson had set out to eliminate from the earlier calculations of Cardinal Gasquet, since Lunn has subsequently confirmed Hamilton Thompson's conclusions and since neither Russell nor anyone else has yet done any work which yields substantially different results, it would seem premature to discard the fruits of their researches. It would be reasonable to say that, if no evidence existed except that of the Ecclesiastical Register, an overall mortality rate among the people of England of at least a third might be expected.

But there is other evidence, and Professor Russell has summarised it faithfully. There is, for instance, the possibility of arriving at an answer through figures for the payment of frank-pledge dues. The value of the calculation is limited since it rests on a narrow statistical base of eighty-four case-histories in Essex, but it is worth noting that it gives an overall mortality of 43%. Court Rolls also provide some evidence, though the principal lesson to be learned from them is the wide variation between different areas. In the Farnham manors the loss between 29 September, 1348, and September, 1350, seems to have been more than 28% but less than 38%, depending on the index figure taken for the ratio between tenants and dependents.[20] A study of the manor of Cuxham in Oxfordshire indicates a death roll of something over two thirds.[21] Similar figures for three manors belonging to Crowland Abbey suggest a rate of 56%.[22] On the other hand, in her analysis based on the Winchester Pipe Roll of eleven widely scattered manors belonging to the Bishop of Winchester, Dr Levett, while venturing no exact figure, could find no evidence to suggest a death rate high enough to disrupt the working of the manors and, in the case of one very large manor, felt that the figure of a third must be over-pessimistic. 'The general impression gained from an attempt to make any such calculations,' she concluded dryly, 'is that they are singularly useless.'[23]

Finally there are the figures derived from inquisitions post-mortem, to which Professor Russell attaches particular importance. Based upon some five hundred such inquisitions he assumes that some 27.3% of the population died during the plague; a figure which would be reduced to 23.6% if allowance were made for the higher mortality among the older people. While admitting the limitations to this approach he concludes, 'Nevertheless, it presents the best evidence available as yet upon the effects of the plague'.[24] 23.6% is far lower than any percentage which can be deduced from the other methods of calculation already mentioned. It might, therefore, be expected that it would be the lowest point in the range of possible death rates. But in his final summing up, Russell puts forward a still lower figure of 20%. 'The reduction of the . . . loss to 20%,' he comments, 'proceeds from better calculation of plague losses, which could take into

account age specific mortality . . . discounting of ordinary
mortality of the three years . . .'[25] Since the 23.6% figure derived
from Russell's favoured inquisitions post-mortem is itself
loaded to take account of age specific mortality this further reduction is hard to accept.

It will be obvious from all this that to draw any conclusion
must be hazardous. Little more than an informed guess is possible. It can safely be assumed that a figure of 45%, equivalent to
the figure of mortality among the clergy, would be the highest
point of any possible bracket. *Pace* Russell, it seems incredible
that the figure could be less than the 23% which he suggested as
the adjusted total derived from inquisitions post-mortem. A half-
way point between these two poles would suggest a death rate of
roughly a third. This would accord reasonably well with the
evidence of the ecclesiastical registers. It is conspicuously lower
than the figure derived from the payment of frankpledge dues
or than most of the manors for which precise statistics happen
to be available but these latter totals are not adjusted to take
account of natural death. It is anyhow not unreasonable to expect
that dramatically bad news would be recorded more enthusiastic-
ally than the humdrum figures of the luckier manors.

As a rough and ready rule-of-thumb, therefore, the statement
that a third of the population died of the Black Death should not
be too misleading. The figure might quite easily be as high as
40% or as low as 30%; it could conceivably be as high as 45%
or as low as 23%. But these are surely the outside limits. On this
basis the approximate total for the dead in England would be 1.4
million. No figure above one million and below 1.8 million would
be astonishing but the nearer that the actual figure approached
the median, the more it would seem to accord with the existing
evidence.

Finally there is the question whether more or less the same proportion holds good for continental Europe. Probably the most
useful observation which one can make on this is that there is no
obvious reason why it should not. The regional variations which
are so evident in England must be, of course, immeasurably
greater in a continent with its wide range of climate, landscape
and racial types. There are certain areas, for instance Tuscany,

where the joint efforts of contemporary chroniclers and modern scholars make it almost certain that the death rate was higher than in England. There are others, for instance Bohemia, where the incidence of the plague was clearly lower. For most of the Continent even the inadequate medieval records which exist in England are lacking or have not yet been exploited—the material for a serious comparison does not exist. But it can still be said that there are few grounds for drawing distinctions between England and the rest of Europe.

Such calculations as have been made for individual countries are not at variance with this somewhat negative assertion. In his careful study of the effects of the Black Death in France, Renouard concluded that the only rule which could safely be put forward was that the overall death rate varied between one-eighth and two-thirds according to the region.[26] Doren has estimated that between 40% and 60% of city dwellers in Italy died but that, in the countryside, the proportion was considerably lower.[27] Neither of these estimates can be translated into a death rate valid for the whole country, a form of calculation which these leading authorities prudently eschewed. In so far as they point in any direction it is towards a figure greater than a third. Certainly they are hard to reconcile with anything substantially lower. But such speculation is unprofitable. To maintain that one European in three died during the period of the Black Death can never be proved but, equally, cannot be wildly far from the truth. Further than that, in the present state of knowledge, one cannot go.

FIFTEEN

The Social and Economic Consequences

One third of a country's population cannot be eliminated over a period of some two and a half years without considerable dislocation to its economy and its social structure. The historian must expect to find conspicuous changes in the life of the English community in the years immediately following the Black Death. At least some trace of the scars will survive into the succeeding decades or even centuries. But exactly what these changes were and how great was their significance has been the subject of bitter and protracted debate. The subject is far from being closed to-day.

The great eighteenth and nineteenth-century historians paid little attention to the Black Death as a force in English history. Hume, in his eight volumes covering the period from the Roman Conquest to the Glorious Revolution of 1688, devoted to the plague one paragraph of sixteen lines.[1] Henry, in twelve volumes, could manage only fourteen lines.[2] Green at least gave it a page and a half and admitted that it had some social consequences but even his treatment was somewhat cursory and he obscurely secreted the passage in a chapter entitled 'The Peasants' Revolt.'[3] Given such conspicuous omissions, it was natural that later historians should celebrate with some exuberance their rediscovery of the Black Death. 'The year of the conception of modern man was the year 1348, the year of the Black Death,' wrote Friedell.[4] It was as significant a phenomenon as the Industrial Revolution, claimed G. M. Trevelyan, though the latter was less striking in its effects since it was not, like the plague, 'a fortuitous obstruction fallen across the river of life and temporarily diverting it.'[5]

The classic exposition of the Black Death's role in England as a social force of the first importance comes from that great medievalist, Thorold Rogers.[6] Many of his conclusions have now been challenged, and challenged with justification, but for breadth of learning, originality of mind and happiness of phrasing he stands far above most of those who have corrected him. 'The effect of the Plague,' he wrote, 'was to introduce a complete revolution in the occupation of the land.' His contention, in grossly over-simplified form, was that commutation, that is to say the substitution of wages and rent in monetary terms for the labour services owed by the villein to the lord, was already well advanced by the time of the Black Death. The sudden disappearance of so high a proportion of the labour force meant that those who already worked for wages were able to demand an increase while those who had not yet achieved this status agitated to commute their services and share in the benefits enjoyed by freemen. If the landlord refused, conditions were peculiarly propitious for the villein to slip away and seek a more amenable master elsewhere.

The landlord was thus in a weak position. Finding himself forced to pay higher wages and obtaining lower prices for his produce because of the reduced demand, he increasingly tended to break up his demesne and let it off for a cash rent to the freemen or villeins of his manor. But he did not succumb without a fight and Parliament came to his rescue with legislation designed to check increased wages and the free movement of labour. The landlords sought to put back the clock and not only to hold on to the relatively few feudal services which still existed but to exact others which had been waived in the period before the Black Death while labour was cheap and plentiful. The result was resentment on the part of the serfs which simmered angrily for thirty years and finally erupted in 1381 in the shape of the Peasants' Revolt.

This sequence of events is plausible and convincing. On the basis of the information available to Thorold Rogers it is, indeed, easy to accept that no more satisfactory pattern of development could have been constructed to bridge the gap between the Black Death and the Peasants' Revolt. His information, however, proved to be far from complete. Subsequent research has demonstrated conclusively that things did not happen according to his tidy scheme. But when it comes to deciding what actually did happen, the impressive unanimity of the historian is significantly less evident. And, within the framework of this problem, the importance to be attached to the Black Death as a factor in the system's disintegration, is far from being definitively established.

Before outlining the counter-arguments which the critics of Thorold Rogers have put forward it would be useful to restate three general considerations, illustrations of which have already been cited at many points but which must constantly be borne in mind if the effects of the Black Death are to be seen in proper perspective. The first of these is that the damage done by the epidemics of bubonic plague in the fourteenth century was cumulative. The epidemic of 1348 was certainly the most devastating and, being the first, by far the best remembered, but further outbreaks occurred in 1361, 1368-9, 1371, 1375, 1390 and 1405.[7]

On the whole these were progressively less violent but the second epidemic of 1361, by any standards other than those of the Black Death, was catastrophic in its dimensions. The progressive depopulation of England which resulted from this sequence of epidemics, as each new generation was attacked before it had made good the losses of the last, was economically and psychologically a depressive quite as dangerous as the holocaust of the Black Death itself. One authority, indeed, has gone so far as to say that the 'most important consequence of the Black Death in fact was simply that the disease was firmly established in England.'[8]

Whenever, therefore, the question arises of the responsibility of the Black Death for any marked change in England—as in the evolution of some new social form or a decline in wealth or population—unless the comparison is strictly between the period before 1348 and the period between 1351 and 1361, then two and not one epidemics have got to be taken into account. If the comparison is made with the state of affairs at some date near the end of the fourteenth century then the problem of responsibilities becomes still more difficult to resolve since three or four epidemics had, by then, taken their toll, as well, of course, as all the other factors which may have contributed to the transformation. It is not uncommon to find that a certain village had, say, fifty-five land holders in 1310 and only thirty in 1377 and for the deduction to be drawn that the Black Death must therefore have been responsible for almost halving the population. It may have been. Almost certainly it was the most important single factor. But in the absence of evidence which will show exactly when and why the drop in numbers took place the contention must remain unproven. Reservations of this kind are still more important when the problem relates not to a fall in population but to a switch from one kind of land-holding to another or to some other social problem.

The second point to remember is that the decline of the economy had already set in well before 1348. No graph could be charted to show the point which the process had by then reached nor was there any consistency between one area and another. But for at least twenty-five years before the Black Death exports, agricultural production, the area of cultivated land and possibly also the population had all been shrinking. In assessing the baleful effects of the Black Death these earlier difficulties must never be forgotten. Continued deterioration in the state of England— and, indeed, of Europe—would have been likely, even if it had never occurred.

Thirdly and finally, the economic impact of the Black Death was to some extent blunted by the fact that England, even though by 1348 there may already have been some decline, was still grossly over-populated. By this it is not meant that the population was greater than the land could support, though this can and has been argued,[9] but merely that the working population had expanded far beyond the work available. In the economic conditions of the fourteenth century this led to chronic under-employment rather than unemployment. Vinogradoff has pointed out that each virgate often had as many as five men working or living on it so that the villein's land could be tended, the service rendered to the lord and a comfortable surplus of labouring capacity still be left unconsumed.[10] Maitland's contention that the landlord was exacting only about half the labour services owed to him, amply confirmed by Miss Levett in her study of the manors of the Bishop of Winchester,[11] is another illustration of this point. It was not through any generosity on the part of the landlord that these services were remitted but rather because there were by so far too many villeins available to do the work that the landlord would have found it quite impossible to employ them all.

This surplus of labour was not confined to the peasants. The lowest computation of the number of priests then available to serve the 8,670 parishes of England is fifteen thousand. For a population of 4.2 million this would give an allowance of one priest for every two hundred and eighty parishioners or more or less every sixty-seven families. Dr Coulton estimates[12] that, even in January, 1349, there must have been three priests surviving to fill every two priestly vacancies. Unless so generous a margin existed it would be impossible to explain how, throughout the plague, almost every vacant benefice was filled within a few weeks.

Bearing these factors in mind, it remains to consider whether the Black Death did indeed bring about as fundamental a revolution in land tenure and social organisation as has been suggested. There is, of course, much that is incontestably true in the thesis. Among the phenomena which Thorold Rogers noted as being particularly relevant were the rise in the level of wages and of prices and the greatly increased mobility of labour. Though regional variations existed, these were variations of degree and duration rather than kind. In every part of England for which evidence exists the tendencies were to be seen.

'The immediate effect of the Plague,' wrote Rogers, 'was to double the wages of labour; in some districts to raise the rate even beyond this.'[13] In Cuxham, a ploughman paid 2s per annum before the plague, earned 7s in 1349-50 and as much as 10s 6d in 1350-51.[14] (These figures are, in fact, not so conspicuously out of line with those of Rogers as at first appears since they ignore certain variations of payment in kind which, if they could be expressed in monetary terms, would certainly go a long way towards reducing the real figures for 1350-51 to about half the apparent total. Two shillings per annum was, of course, an inconsiderable element in the ploughman's total remuneration.) In Teddington and Paddington wages were doubled in the first year.[15] Rogers's own figures show that a thresher who had been paid 2½d a day in 1348 earned 6d in 1349 and 4½d in 1350, while a mower received 5d an acre in 1348 but increased this to 9d in 1349 and 1350.[16]

All this supports the thesis that wages more or less doubled. But not all evidence points to so sharp an increase. In her examination of the manors of the Bishop of Winchester, Professor Levett found that in some, though by no means all cases, wages rose by between a quarter and a third[17] while Lord Beveridge was able to detect only minor increases in another group of the

Bishop's manors[18] but a jump of more than 75 % in the wages paid on the estate of the Abbot of Westminster.[19] But even though Rogers's estimate may be on the generous side as an average for the whole of England, it is clear that wages rose rapidly and substantially and imposed a heavy burden on any landlord who depended largely on paid labour to farm his demesne.

Professor Rogers and the other proponents of his theories are also undoubtedly right in saying that prices of agricultural products fell steeply during and directly after the Black Death; thus making still more troublesome the life of the landlord. Lack of demand was, of course, the prime cause. Knighton's complaint that 'a man could buy for half a mark a horse which formerly had been worth forty shillings,'[20] has already been mentioned. Rogers has shown that oxen which fetched 13s 7d in 1347 and 10s 6d in 1348, were sold for only 6s 8d in 1349. A cow fell in value from 9s to 6s 6d and a sheep from 2s 2d in 1347 to 1s 5d in 1348, 1s 4d in 1349 and 1s 3d in 1350. Poultry seem more or less to have maintained their prices and corn did reasonably well because of a poor harvest in 1349, but the price of wool was lower than at any other time in the fourteenth century. Against this, the cost of manufactured products, many of which the landlord would have had to buy, tended to rise steeply because of difficulties of transport and of the death rate among the skilled artisans who, unlike the agricultural workers, had no pool of surplus labour ready to fill the gaps. Every one knew how to cut hay but few indeed were competent to make a nail. The price of canvas rose from 2s 3½d in 1347 to 2s 9d in 1349 and 4s 3d in 1350; a bushel of salt which cost 4⅝d in 1347 could not be bought for less than 1s 2d in 1350; iron jumped from 8s 6d to over a pound for twenty-five pieces.[21]

The landlord was to some extent sheltered against these difficulties by the extra income which accrued in 1349 and 1350 from the entry fines levied on the estates of the dead before the heir was allowed to take them over and the cattle which were collected as heriots, a form of death duty paid in kind. But this was a once-and-for-all increment and, anyway, often had to be waived in cases where no heir survived or the survivor could not afford to pay the fine. Heriots, indeed, could be an embarrassment. In Farnham where fines rose from under £20 to over £100, the reeve was forced to take on extra meadows and engage more labour because it was impossible to dispose of the influx of cattle at a reasonable price.[22]

Professor Rogers's third point is no less valid. There was undoubtedly greater mobility of labour during and directly after the Black Death and any landlord unready to make concessions to his tenants might well find that they had vanished to seek a kindlier master. The case of the tenants of the manor of Woodeaton who 'would have departed had not Brother Nicholas of Upton, then Abbot, made an agreement with them . . .'[23] has already been mentioned. At Forncett, in the generation succeeding the Black Death, over half the customers' tenements and a quarter of the sokemen's reverted to the lady of the manor because of the death or flight of the sitting tenants and were subsequently relet at a money rent.[24] Some of those who fled turned up a little later on neighbouring manors, others disappeared altogether, either to farms in more distant parts or, perhaps, to make a new life in the rapidly expanding cloth trade.[25] A Lincolnshire ploughman refused to serve except by the day and unless he had fresh meat instead of salt. When he could not get what he wanted, his retort was to disappear and offer his services elsewhere.[26]

It would be pointless to multiply such instances. The proof that labour was on the move is provided by the energetic efforts which the Government made to check it. The Ordinance of Labourers of 1349 and the subsequent Statute of Labourers in 1351 were, *inter alia*, a direct attempt to prevent workmen transferring their loyalties from one employer to another. Except for the application of the statute to freeman as well as bond there was nothing sensationally new in this. The most interesting feature, indeed, was that it had proved necessary to pass a statute at all and, incidentally, to set up complicated and expensive administrative machinery to enforce it. 'The statute of labourers,' wrote Dr Putnam, 'must be regarded not as having created a new system or a new set of economic relations but as affording proof that radical changes had occurred, ushering in a new era.'[27] Perhaps the most radical of these changes was the new desire, even determination on the part of the medieval labourer to have a say in deciding his terms of employment and to seek his fortune elsewhere if such a right were denied him.

It is no more possible to dispute that these phenomena existed than it is to doubt Rogers's contention that the landlord—unable to hire labour except at greatly increased wages, unable to get a good price for his products or to buy what he needed for the farm except at exorbitant cost; unable to enforce his manorial rights because the villeins fled when he attempted to—was sorely tempted to abandon the struggle altogether. His remedy was to let off the demesne to the tenants for a cash rent in units small enough for them to farm themselves. To take only one example; in the bailiwick of Clare, on one manor at least, all new leases made after 1349 were for money without labour, leases of the demesne lands became common from 1360 onwards and, by 1380, the greater part of the demesne had passed out of the lord's possession.[28] The Black Death introduced a situation in which land was plentiful and labour scarce. The scales were thereby tipped against the land owner.

It is the peculiar virtue of English society that it contrives permanently to remain in a state of transition; no sooner has it crossed one bridge than it is off on its uncertain course across the next one. Sometimes, indeed, it tries to cross two at once. It would thus be hazardous to argue that England in the fourteenth century was more conspicuously in transition than at any other period but, certainly, it would be hard to find an age in which the change was more fundamental. The pattern of several centuries was breaking up; not only the pattern of society but the set of men's minds as well.

'Increasingly, the really significant distinction,' wrote Dr McKisack, 'is less between free and servile than between winner and waster, between the man whose fortunes are on the upgrade, whose descendants may well swell the ranks of the yeomanry and gentry of a later age, and the man whom economic pressure or lack of enterprise are driving downwards.'[29]

The moment, therefore, was one of great fluidity. In such circumstances even a mild contretemps can produce disproportionately sharp reactions. The consequences of any more severe shock are likely to be intensely violent. Few shocks can have been more violent than that caused by the Black Death in fourteenth-century Europe.

We can thus safely agree with Rogers that the Black Death must have led to important changes in the social and economic structure of the country. We can accept too the existence of the phenomena which Rogers noted and on which he based his thesis. But now begin the qualifications. In medieval history, it seems, it is not the exception which proves the rule but the rule which generates the exception. A useful principle can be established which may, in memory of that great iconoclast, Dr A. E. Levett, be described as Levett's Law: 'The enunciation of any authoritative statement shall immediately be countered by the accumulation of evidence leading to its contradiction.' Once a sufficient body of such material has been compiled then, of course, a new statement is deemed to have been advanced and the work of destruction can change its target.

So, as soon as Thorold Rogers had won general acceptance for his thesis of a Black Death causing profound social changes and leading directly to the Peasants' Revolt, evidence began to be elicited to prove him wrong. He had argued, it will be remembered, that commutation was already far advanced by 1348, and that it was above all the efforts of the landlords to reverse this process which led to social unrest. As a first stage in the demolition of his theories it was demonstrated that commutation had, in fact, made little progress before the plague.

Mr Page[30] argued that there were not enough freemen available to do the commuted labours. Even if there had been, the amount of money in circulation would not have been sufficient to pay them or to allow the former villeins to pay their rents in cash. He analysed the sources of labour on a group of eighty-one manors spread over twenty counties for the period 1325-1350. On rather over half these, villeins did nearly all the work on the lord's demesne, on twenty-two of them they did about half, on nine an insignificant amount and on only six had predial service been abolished altogether. A similar analysis for the year 1380 of a group of one hundred and twenty-six manors, of which fifty-five had been included in the earlier examination, showed that the villeins did all the work on only twenty-two, about half the work on twenty-five, an insignificant amount on thirty-nine while on the remaining forty no predial services at all survived. A dramatic transformation had clearly taken place in the intervening years, vindication of the statement made by Pollock and Maitland: '. . . it was the Black Death which, by destroying nearly half the population while leaving the available capital and the medium of

exchange as great as ever, hastened the transition from a system of barter to a system of money payments.'[31] Furthermore,
on no manor did Page find evidence of any villein being held to
service after the Black Death if this had not also been the case
before.

Studies of this nature certainly contradicted the conclusions of
Thorold Rogers but tended to increase rather than diminish the
significance of the Black Death. Page's conclusions were not for
long suffered to remain uncontested. Sir Keith Feiling conducted
a closer analysis of one of his selected manors and pointed out that
several of the leases for money rent on which Page based his case
were in fact short term only and were later replaced by leases
based on the old scale of rents and services. Several categories of
manorial services were definitely more onerous by 1362 than they
had been before the Black Death. Two and a half *opera minuta*
were commuted in 1341-2, 181 in 1353-4, 285$\frac{1}{2}$ in 1354-5, 77 in
1358-9 and none in 1362-3.[32] On this manor at least, therefore, the
pattern was not one of a process of commutation initiated by the
Black Death and swelling triumphantly to its near completion at
the end of the fourteenth century, but rather of a false start directly
after the Black Death which was soon checked and reversed by
the victorious landlord.

The next salvo came from Dr H. L. Gray[33] whose techniques
have been much criticised but whose signally useful if modest
contribution was to rationalise and codify the doctrine that any
generalisation was futile if it professed to apply to the whole of
England. Based on his study of inquisitions post-mortem he
claimed that, in the North and West of England, services had
nearly always been commuted before the Black Death; in
the South and East full or, at least, a large number of services
were still exacted in about half the manors while in Kent,
serfdom had died out at a much earlier date. It is hardly necessary
to point out how quickly examples were forthcoming to show
that generalisations were hardly more valid when applied to areas
such as the North or the South, than to the country as a whole.
But as a rough and ready rule, the value of Gray's analysis remains considerable.

Dr Levett has provided the weightiest evidence against the
presentation of the Black Death as a watershed in English social
history. She has shown that, on many of the manors of the Bishop

of Winchester, commutation was hardly known before the Black Death and that there was remarkably little change introduced in the years immediately afterwards. When William of Wykeham did begin to indulge in substantial commutation of services against money payments it was more to raise money for his projects at New College and Winchester than because of pressures generated by the Black Death.[34] Her statistics are impressive but against them it can be contended that the experience of manors belonging to rich, powerful and conservative churchmen need not necessarily be applicable even to other manors in the same neighbourhood, let alone to the country as a whole.

Nor did her findings apply even to all the manors of the Bishop. A companion study of the Manor of Witney reaches remarkably different conclusions.[35] In that manor, two thirds of the population died, much land was thrown on to the lord's hands and had to be let on new terms and for a money rent, the landlord gradually abandoned the struggle to farm the demesne and, having no further use for labour services, willingly commuted them for cash. The number of villeins employed at harvest-time dwindled from one hundred and twenty-one in 1348 to twenty-eight in 1350, rose again to forty-two in 1352, but by 1354 had only reached forty-eight. All new tenants were excused tenurial labour and the whole system quickly withered away. Similarly, on the estates of Ramsey Abbey, the Black Death heralded the introduction of a new style of rent involving a larger cash payment, but the disappearance of all, or virtually all villein services.[36]

To take another example in which the landlord similarly enjoyed enough wealth and influence to ride almost any economic storm; at Cuxham, a manor of the College of Merton, prior to the Black Death well over half the work on the demesne was done by the *famuli*, labour attached to the manor and living within its compound. Two thirds of the rest was done by customary labour and only a third by hired labour. The proportion of work done by the wage-earner actually dropped in the seventy years between 1276 and 1347—the period during which, according to Thorold Rogers, commutation was rapidly gaining ground all over England. After the Black Death labour services virtually ceased and work done by the *famuli* was halved. Total work done on the demesne was reduced but the cost of hired labour nevertheless trebled. In 1361, Merton College decided to lease the manor rather than continue to run it at a loss.[37]

In some parts of England, therefore, the Black Death was a sharp stimulus towards rapid and lasting commutation of manorial services, in others it gave rise to much commutation but the landlords were able to check the process and more or less restore the *status quo ante*, in yet others it had little perceptible effect on the manorial structure and, finally, in a few it impelled the landlords into a reaction which sought to resurrect labour services that had long been suffered to fall into disuse. The more prosperous and stable manors were the least affected; where the land was poor, the landlord ineffective, or the disease raged with especial violence, then the consequence was likely to be a rapid growth in commutation. It would be impossible to estimate in how many cases this development was something entirely new and in how many the process was already under way. Professor Postan has argued in favour of a rapid move towards commutation in the twelfth century which slackened or even went into reverse in the course of the thirteenth.[38] However that may be, it is reasonable to contend that commutation was well-known in most parts of England before 1348, and that the Black Death did no more than accelerate, though often violently accelerate, an established and, in the long run, inevitable progress.

What of the other economic and social effects of the Black Death which Thorold Rogers maintained did so much radically to change the manorial system and lead towards the Peasants' Revolt? Wages and prices of manufactured goods certainly rose sharply after the Black Death but this rise was not maintained. Nor was the fall in the value of agricultural products. Almost all the examples cited earlier to illustrate the dramatic effects of the plague can also be used to show how quickly the effects passed. But for the most part they did not pass altogether. Particularly in the case of wages a very real advantage was won and retained by the labourer in almost every part of England. The ploughman of Cuxham whose pay had risen from 2s to 10s 6d was still earning 6s 3d in 1351-2 and 7s 6d over the period 1353-9. In Teddington and Paddington, wages fell back sharply in 1351-2 but remained well above the figure for before the plague.[39] Thorold Rogers's thresher, whose wage averaged 3½d in the first half of the century, earned 4¼d in the second half while the carpenter's wages rose from 3⅛d to 4⅝d.[40] The rise in the cost of living took away some of the wage earner's advantage but his rent, probably the most important item in his budget, if it increased at all, certainly did not do so as substantially as his income. His net advantage was almost always considerable. *Post hoc* is not necessarily *propter hoc*, but it would be ultra-cautious not to admit that the Black Death was a major factor in the process.

Prices of agricultural produce seem on the whole to have more than regained their level within a year or two of the end of the plague, though they lagged behind the index for wages. Taking the two ten-year periods of 1341 to 1350 and 1351 to 1360, wheat, barley and other grains rose by up to 30%, but the price of wool dropped slightly and live-stock varied so wildly as to make any deduction virtually impossible. Oxen fetched about 15% more but cows about 3% less; sheep substantially more, pigs and cart-horses slightly less; pullets and ducks more but hens, geese and cocks less. The price of manufactured goods, on the other hand, dropped back a little from the abnormally high level of the years of the Black Death and immediately after but still remained well above the pre-plague average. Salt, which cost 6¼d the bushel the decade before the plague, cost 10½d between 1351 and 1360. Iron varied according to type but all types cost more and some increased threefold. Clouts almost doubled in price while canvas leapt from 2s 5d for the dozen ells to 6s 5d.

In so far then as it can be assumed that the Black Death was primarily responsible for the altogether exceptional trend of wages and prices between 1340 and 1360—and such an assumption can surely be safely made—then it is clear that it did the landlord little good and much harm. Even if he managed to maintain agricultural production at its previous level, he could expect to receive little more and perhaps even less for his produce while having to pay substantially more for his labour and his imported articles. Wool, by far the most important crop produced for sale rather than consumption, actually brought the farmer a smaller return in the decades after the plague than before 1349. The blow was not economically devastating except, perhaps, in 1350 and 1351 and, during these years there was usually extra income from other sources to sustain the landlord. But it was certainly painful enough to provide a powerful disincentive to anyone wondering whether or not to carry on the farming of his demesne.

Thorold Rogers's argument rested above all on the hypothesis that the Black Death so far reduced the population that those who remained were placed in an immeasurably stronger position when it came to bargaining with an employer. In the short term—that is to say in 1349, 1350 and 1351—this was of course true. If a third of the peasants of a given area disappeared within a few months then, whatever the reserves in labour, there was bound to be serious dislocation. But provided the labour reserve was great enough—and it has already been argued that it was substantial[41]—then an adjustment of resources to needs was bound, in time, to put the matter right. In some areas the process of adjustment would be relatively simple; in others, where the Black Death did its worst damage, it would be painful and protracted. But in the end it would be done.

Again and again in the patchwork of horror stories which composes our knowledge of the Black Death one of the most striking features has been the speed of recovery shown by the medieval community. In all the manors of the Bishop of Winchester which she studied, Dr Levett found only a very few where tenements remained vacant for more than a few years. On the estates of Crowland Abbey, where eighty-eight holdings were left empty, all but nine of these were quickly taken up; not by peasants from other villages who might have deserted land elsewhere and so left another gap to fill but by people with names already known on the manor who, one must presume, were landless residents before the plague. The estate of the Abbey, in fact, had sufficient surplus of man-power to fill even the huge vacuum left by the plague. At Cuxham, nine out of thirteen half-virgates were still vacant by March, 1352 and in this case recourse was had to importing tenants from outside the manor. Within another three years all the vacancies were filled. Yet it would be a mistake to suggest that this was an easy or painless process, or that all areas recovered so completely. At Standon, for instance, one of the worst affected manors of the Earls of March, many tenements stood unoccupied until 1370. Even in the less depopulated areas the balance between work to be done and labour available was bound to be more precarious than in the past. England had consumed her fat and it was going to be far more difficult for it to recover a second time if any fresh strain were imposed.

Such a strain was to be imposed with the second epidemic of bubonic plague in 1361. In the meantime, however, the relatively light incidence of the Black Death among the generations most likely to bear children coupled with the new wealth and economic opportunities released by the great mortality, had produced an unusually high birth rate in the intervening years. A monk of Malmesbury, it is true, remarked that, 'the women who survived remained for the most part barren during several years,'[42] but the evidence for the statement is obscure. At the most it can be taken as applying only to the period at the end of and immediately after the epidemic when the sense of shock was still in the forefront of men's minds and they might have deemed procreation offensive to the Almighty. Obviously by 1361, the children of the post-plague years were not yet competent to undertake the work done by their deceased uncles and cousins, but numerically at least the recovery had begun. It was only after 1360, and still more in the last quarter of the fourteenth century, that depopulation began substantially to change the face of England.

Another point to which Thorold Rogers attached particular importance was the ease with which the peasant could escape from his manor in the chaotic conditions of the English countryside in 1349 and 1350. This ever-present if unvoiced threat must have made the landlord far more amenable to the peasants' pleas for better conditions of work. It is only fair to say, however, that on most manors there was little to stop a villein escaping even before 1349. He probably had only to step over a brook or cross some invisible demarcation line to put himself beyond the reach of his master except through complicated and usually expensive legal processes. Given that the landlord was likely to have had more than enough labour on his estate already, it was unlikely that he would pursue his recreant villein with any vigour. 'It cannot be urged too often,' wrote Vinogradoff,[43] 'that the real guarantee against a dispersion of the peasantry lay in the general fairness of the conditions in which it was placed.'

After the Black Death many villeins, viewing enviously the high wages earned by those no longer bound to render predial services, began to think that the conditions in which they were placed were no longer generally fair. Rogers is therefore surely justified in his belief that the Black Death was a stimulus towards greater mobility of labour and hence towards the disintegration of the manorial system. But the legislation which this new mobility provoked to counter it went far towards nullifying this result. For a long time it was accepted doctrine that the Ordinance of Labourers and the subsequent Statute of Labourers were dead letters from the start; ignored by the labourers and treated with indifference or contempt even by the employers themselves. Knighton, with his categoric statement: 'Labourers were so elated and contentious that they did not pay attention to the command of the King,'[44] was perhaps the father of this thesis, but the vision of the sturdy British peasant standing up stoutly to any interference with his liberties by wicked barons or cold-hearted bureaucrats was calculated to appeal irresistibly to any Whiggish historian. The laws should have failed and therefore they did fail.

It is hard to reconcile this sympathetic doctrine with the facts. The object of the statutes was to pin wages and prices as closely as possible to a pre-plague figure and thus to check the inflation that existed in the England of 1349-51. The Government realised that this could never be achieved so long as labourers were free to move from one employer to another in search of higher wages and so long as employers were free to woo away labourers from their neighbours with advantageous offers. By restricting the right of an employee to leave his place of work, by compelling him to accept work when it was offered him, by forbidding the employer to offer wages greater than those paid three years before, by making illegal the gift of alms to the able-bodied unemployed and, finally, by fixing the prices which butchers, bakers and fishmongers could charge their customers, they hoped to re-create the conditions that pertained before the plague and maintain them for ever. The statute of 1351 took this one stage farther by codifying the wages of labourers and artisans.

This was, of course, a hopeless quest. But, though any analogy to the twentieth century would be ridiculous, it must be admitted that, as prices and incomes policies go, the fourteenth-century freeze was remarkably successful. Between 1349 and 1359, six hundred and seventy-one men were appointed to enforce the statutes. Though the bulk of the prosecutions were inevitably of offending peasants, the employer did not escape entirely. Dr Putnam records cases of one employer prosecuted for 'eloigning' the servant of another with an offer of high wages, a rector prosecuted for paying his household servants too much and a reeve for hiring reapers in a public place at an illegal rate.[45] On the whole the statutes were not imposed with severity, whether against employer or employed. Imprisonment was extremely rare and fines for the most part moderate. The result is self-evident. Within a few years wages and prices had fallen back; not indeed to the pre-plague level, but at least to a point well below their maximum. Governmental action cannot be given all the credit for this; it is probable that there would anyhow have been a reaction once the immediate shock of the Black Death had worn off. But equally it seems unreasonable to dismiss as a total failure legislation which, in fact, achieved most of what it set out to do.

In defence of the statutes it can be said that, though loaded heavily against the peasant, they were not conceived solely as instruments for his repression. Certainly, in part, they were inspired by the fear that labour would get out of control but also they reflected a genuine wish to prevent the wealthy land-owner or industrialist drawing away labour from his weaker rival.[46] They can, therefore, be presented as seeking to protect, if not the poor, at least the not-so-rich. But any legislation which imposes a maximum but no minimum wage and which expects the baker— whose interest it is to see prices rise—and the farmer—whose interest it is to see wages fall—to respond in the same way to legislation suggesting that both prices and wages should remain as they were, must inevitably discriminate against the poorer classes. The laws may not have been intended to repress but they were administered largely by the land-owners in their own interests. Inevitably it was the labourer who lost. For the most part the Statutes did not operate so as to make the labourer worse off than he had been before, but they cut off a line of advance towards a new prosperity which had been opened by the plague. The fact that they were largely successful was an important factor in the compound of national issues and local grievances which was eventually to give rise to the Peasants' Revolt.

Can it be said therefore, in schoolboy phrase, that the Black Death 'caused' the Peasants' Revolt? The classic thesis that it was the reversal of a far-advanced trend towards commutation which provoked resistance among the peasants must in part at least be rejected. If, on manors as numerous and as scattered as those of the Bishop of Winchester, Dr Levett can find 'absolutely no sound evidence for retrogression or greater severity in exacting services after 1349,'[47] then no generalisation which assumes the existence of such retrogression can be wholly valid. Certainly the same is not true in every part of England: there were cases in which peasants were forced back into a servile status from which they had previously escaped. Such cases were undoubtedly resented. But in sum there is no reason to think that these constituted a major, let alone the major, factor in instigating the uprising.

What then did cause a rebellion as determined and as widespread as that of 1381? Petit-Dutaillis, who may be said to have spear-headed the attack on the established point of view, considered that it was a compound of irritating feudal burdens, mainly in the form of financial exactions, and the clumsy tax policy of the royal advisers. 'The contradiction which existed between their legal state and their economic advancement was evidently a source of daily exasperation.'[48] Professor Hilton, who saw the genesis of the Peasants' Revolt at the very beginning of the thirteenth century, has analysed the factors which led to unrest.[49] Many of them, it will be obvious, were active irritants long before 1348. The undue conservatism of the landlord who sought to preserve the irritating frills as well as the essential spirit of the manorial system, the denial to the peasant of the right to dispose of his chattels, the fact that prices rose faster than wages, the resentment of the villein who saw his free neighbour exploiting the new circumstances to the full, the inequity and uneven incidence of the poll taxes, the abduction of peasants by labour-hungry landlords, the curbs on liberty of action imposed by the new legislation: these were the elements which finally provoked explosion.

But because the Black Death was not an immediate cause it does not follow that it should not bear a large share of the responsibility. If there had been no plague it is arguable that the circumstances which so disturbed society in 1381 would eventually have arisen. The break-down of the structure of a society can never be painless and, by the second half of the fourteenth century, the disintegration of the manorial system was inevitable and already well advanced. But the Black Death immeasurably aided the process; exacerbated existing grievances, heightened contradictions, made economic nonsense of what previously had been a situation difficult but still viable.

'The Black Death did not, in any strictly economic sense, cause the Peasants' Revolt or the break-down of villeinage, but it gave birth, in many cases, to a smouldering feeling of discontent, an inarticulate desire for change, which found its outlet in the rising of 1381.'[50] Dr Levett's cautious judgement may be termed the lowest statement of the Black Death's claim to have inspired the social unrest of the later fourteenth century. It cannot be said, in short, that the Black Death directly caused the Peasants' Revolt, nor can it be said with certainty that, but for the Black Death, the Peasants' Revolt would never have taken place. But what can be asserted with some confidence is that, if there had been no Black Death, tension and bitterness would never have risen by 1381 to the level that it did.

'We must really not raise the plague to the dignity of a constant, economic force.'[51] Vinogradoff's magisterial warning should be written in scarlet above the desk of every historian dealing with the fourteenth century. But we must really not lower the plague to the level of an isolated phenomenon having no significant influence on the development of the country. The more extravagant claims of its champions may be discounted but so also may the excessive denigration of those who sought to cut it down to size. The Black Death did not initiate any major social or economic trend but it accelerated and modified—sometimes drastically—those which already existed. There is at the moment, particularly in Germany, a sentiment in favour of reinstating the Black Death as a major originating factor in its own right.[52] Such a counter-reaction is desirable and, indeed, overdue. It should not be taken too far but it will not have been taken far enough until it is generally accepted that the Black Death was a catalytic element of the first order, profoundly modifying the economic and social forces on which it operated. Without it the history of England and of Europe in the second half of the fourteenth century would have been very different.

SIXTEEN

Education, Agriculture and Architecture

Such modifications of the social structure of the country were bound to find their reflection in almost every sphere of human activity. There can have been very little in English life which survived the Black Death wholly unchanged, though in some fields the changes were at first almost imperceptible and only gradually revealed their true significance.

The world of education, through its dependence on a comparatively small group of learned men of whom the most powerful and distinguished were often also among the oldest, was peculiarly sensitive to the impact of the plague. Mortality among men of learning had been calamitously high. Four of Europe's thirty universities vanished in the middle of the fourteenth century: no one can be sure that the Black Death was responsible but it would be over-cautious to deny that it must have played a part.[1] Arezzo ceased to exist a few years later; Siena closed for several years. The chancellor of Oxford petitioned the King 'showing that the university is ruined and enfeebled by the pestilence and other causes, so that its estate can hardly be maintained or protected.' The students of Avignon addressed the Pope: '. . . at a time when the university body of your studium . . . is deprived of all lectures, since the whole number has been left desolate by the death from pestilence of doctors, licentiates, bachelors and students . . .'

Into this vacuum there was ample scope for new ideas and doctrines to infiltrate. In England one important by-product, caused in part at least by the shortage of people qualified to teach in French after the Black Death, was the growth of education in the vernacular and of translation from Latin direct to English:[2] John Trevisa said of the old system—the translation of Latin to French:

'Thys manere was moche y-used tofore the furste moreyn, and ys sethe somdel ychanged. For Johan Cornwall, a mayster of gramere, chayngede the lore in gramer-scole, and construccion of Freynsch into Englysch; and Richard Pencrych lurnede that manere teching of hym, and other men of Pencrych, so that now, the yere of our Lord a thousand, three hundred foure score and fyve . . . in all the grammer-scoles of England childern leaveth Frensch and construeth and lurneth ye Englysch and habbeth thereby avauntage in on syde and desvauntage yn another. Their avauntage ys that they lurneth gramer in lesse tyme than childern were i-woned to doo; desavauntage ys that now childern of gramer-scole canneth na more Frensche then can thir lift heale, and that is harme for them an they schulle passe the see and travaille in straunge landes . . .'

Viewing the history of the English language and its literature over the last six hundred years there are few who would deny that John Cornwall, mayster of gramere, deserved well of his country. The disadvantage to which Trevisa refers still exists to trouble us but the fruits of Cornwall's reform outweigh immeasurably the gulf which it placed between this island and mainland Europe. It would, of course, be absurd to attribute to any individual or to the Black Death itself the full responsibility for a change which had already started even before 1348 and, in the end, would inevitably have carried all before it. But it would also be a mistake to discount unduly the importance of the Black Death in removing so many of those who would have been a barrier to reform and in making it, in purely practical terms, far more difficult to carry on along the old path.

John Cornwall's innovation was more fundamental than is suggested in Trevisa's chronicle. For the growth of a literature in the vernacular was bound, in the end, to mean the disappearance of Latin as a medium of communication. It took an unconscionably long time a-dying; vestigial relics are, indeed, still said to linger on in England to-day in certain of the more antique seats of learning. But its monopoly was broken. English arose; the symbol of a new nationalism, to take its place in the law courts as the instrument for the transaction of business and for the conduct of relationships even in the most polite society. It was nationalism that dictated the use of English rather than the use of English which created nationalism, but the two fostered each other and grew side by side. Neither the growth of a national language nor of a national spirit can be said to be a uniquely English phenomenon. The sort of generalisation with which we are dealing here could be applied, *mutatis mutandis*, to what we now mean by France, Italy or Germany. But nowhere else was the evolution so pronounced or the relevance of the Black Death so clearly marked.

'We must not think,' wrote Dr Pantin,[3] 'that "nationalism" was something invented at the Renaissance or even in the later middle ages . . . since the eleventh century there had been highly organized "national" states and deep political and racial divisions and rivalries and antipathies . . .' One must not give the Black Death too much prominence in an evolution which has edged forward fitfully over many centuries, yet it would be quite as foolish to ignore its role and it is surely permissible, too, to see its by-products in the field of learning as one of the more decisive catalytic factors. In England too, it was more immediately apparent that the weakening of the international language was a blow to the universal Church. It would be a grotesque over-statement to claim that, if the English had continued to speak French and write Latin, there would have been no Reformation, but, like most over-statements, it would contain some elements of truth.

In the long run the English Universities had no cause to regret the temporary havoc which the plague caused in their workings. The shortage of clergy and lay clerks was so conspicuous that the provision of replacements became an urgent need. At Cambridge the reaction was swift. Trinity Hall, Gonville Hall and Corpus Christi were all founded as a consequence, Corpus Christi, at least, as a direct consequence of the Black Death. In the deed of 6 February, 1350, by which Bishop Bateman established Trinity Hall it was specifically laid down that the purpose of the new college was to make good the appalling losses which the clergy in England and, in particular, East Anglia had suffered. The motives for founding Corpus Christi were slightly less altruistic. The members of the trade guilds found that, with the shortage of clergy after the plague, it cost them too much to have masses said for their departed members. By establishing a college they calculated that they would acquire a plentiful supply of cheap labour among the students. So deep a scar did the plague leave that even in 1441, at the foundation of King's, the statutes, though in general terms, reiterated a reference to the need to repair the ravages of a century before.[4]

Oxford was somewhat slower off the mark. It took ten years and a second attack of the plague to induce Simon Islip, Archbishop of Canterbury, to follow Bishop Bateman's lead. '. . . I, Simon . . . in view of the fact that in particular those who are truly learned and accomplished in every kind of learning have been largely exterminated in the epidemics, and that, because of the lack of opportunity, very few are coming forward at present to carry on such studies . . .' heartily support a gift of money made to my 'new college of Canterbury at Oxford.'[5]

Similarly, a few years later, William of Wykeham, wishing to cure 'the general disease of the clerical army, which we have observed to be grievously wounded owing to the small number of the clergy, as a result of pestilences, wars and other miseries of the world'[6] founded New College to repair the deficiency. But New College owed more to the Black Death than the inspiration for its creation. According to tradition, backed by Thorold Rogers,[7] New College garden was the site of Oxford's largest plague pit, an area formerly covered by houses but depopulated by the epidemic and converted to its grisly purpose. The ill wind also blew good to Merton College since the sharp drop in population allowed it to buy, at a bargain price, almost all the land between the City Wall and St Frideswides; an investment whose increasing value must have done much to solace future generations for the tribulations of their ancestors.

It would have been extraordinary if the striking changes which the Black Death had helped to evolve in the relationship between landlord and tenant had not produced perceptible results in the practice of agriculture and even the appearance of the English countryside. The crucial consequence of the epidemic was that much land fell free and that the lord not only did not wish to farm it himself but was often anxious to divest himself even of that part of the land which had formed part of his demesne before 1348. The tenements of those who died and left no heir were therefore available for distribution among those who remained.

Sometimes such tenements might be taken up by immigrants who had sickened of their own, less fertile holdings and let the wilderness take over its own again. But more often the lands of the deceased were carved up among the surviving tenants of the village.

Each tenant, therefore, was likely to have a larger holding than before and, in the fluid conditions which prevailed after the plague, these could be organised into more coherent and viable blocks than had been possible under the old pattern of cultivation. The tendency was reinforced where the landlord alienated his demesne. Once a tenant was established in possession of a coherent parcel of land, then it was inevitable that he would seek to demarcate it more clearly and organise his different activities on a more workable basis. It would be wrong to speak of any dramatic and sudden switch; it took generations to transform the face of the countryside. But the hedged fields of England can plausibly be argued to have had their genesis in the aftermath of the Black Death and though such changes would, in the long run, have been inevitable, their evolution might otherwise have followed a distinct and far more protracted oath.

Text-books have often nurtured the tradition that another consequence of the Black Death was a wide-spread switch from arable to sheep-farming. The logic of such a development is clear. As a result of the plague, labour was scarce and dear—what more natural than to switch from labour-intensive crops to sheep which called for a minimum of skilled attention? But because something could reasonably be expected to have happened, it does not follow that it did. In fact there is little evidence to show that there was any movement to pasture farming, none to show that the movement was general throughout the country. The acreage under plough certainly dropped but this was no more than a symptom of the retreat from the less profitable marginal lands which was already marked before the plague. There is no corresponding increase in wool production to set against this trend: on the contrary, the third quarter of the fourteenth century is one of diminished demand for wool and of stagnation or even decline in English sheep-farming.[8] The great swing to sheep, with its concomitants of vastly increased national prosperity and the harsh social policy of enclosure, was checked rather than advanced by the Black Death.

Another field in which the significance of the Black Death seems more significant in legend than in reality is that of architecture. The skilled masons capable of executing the fine traceries and, still more, the figure sculpture of the Decorated period were, it is contended, almost wiped out by the plague. Those who were left were too much in demand, too pressed for time, to be able to use their talents to the full. The new generation of masons, artisans rather than artists, were affected by the new mobility of labour which was so marked a feature of the post-plague period. Forced to work in a variety of stones, most of them unfamiliar, it was inevitable that the workmen should opt for less complicated and ambitious techniques.[9] The result was a sharp fall in standards. Prior and Gardner write with disdain of the 'same stereotyped monotony, the same continuous decline in the skill of execution, the same obvious diminution of interest in the craft of the artificer' which were to be found in the detailed work of the period.[10]

There is, of course, something in this argument. Without doubt many skilled craftsmen died during the plague and were never replaced. With them died one of the glories of English religious architecture. There can be no absolute standard of beauty but most people would probably agree that York Minster would be more perfect a building if work had begun ten or twenty years before it did. As it was, work came to a sudden stop on the almost completed west front and nave. The choir had not yet been begun and no further progress was made till 1361. For its construction the old plans were scrapped and the Decorated style replaced by the formal stiffness of the Perpendicular. One reason at least for this must have been the technical impossibility of continuing to build a Decorated church when so many of the more experienced masons were dead.[11]

But can a trend which led directly to the towers of Worcester, the west front of Beverley Minster or the nave of Canterbury, possibly be cited as evidence of inferior artistry? To suggest that the Perpendicular style was no more than a degenerate variant on the traditional Decorated would show a derisory misunderstanding of one of the noblest schools of English architecture. Nor should the significance of the Black Death be over-stated in an artistic revolution which had started twenty years before and which the calamities of the mid-fourteenth century checked but could not extinguish. The transept and choir of Gloucester, the cradle of Perpendicular, were completed in 1332 and, though the Black Death introduced economic and social factors which diffused the new fashion more widely, it would be misleading to suggest that those were of prime importance. 'Perpendicular,' wrote Mr Harvey, 'was not the outcome of poverty and failure, but of riches and success. Only to a comparatively slight extent was its course changed by the coming of the Black Death, which did but accelerate a movement already in being.'[12]

SEVENTEEN

The Effects on the Church
and Man's Mind

'The plague not only depopulates and kills, it gnaws the moral stamina and frequently destroys it entirely; thus the sudden demoralisation of Roman society from the period of Mark Antony may be explained by the Oriental plague . . . In such epidemics the best were invariably carried off and the survivors deteriorated morally.

'Times of plague are always those in which the bestial and diabolical side of human nature gains the upper hand. Nor is it necessary to be superstitious or even pious to look upon great plagues as a conflict of the terrestrial forces with the development of mankind . . .'

It may reasonably be felt that Niebuhr was pitching it a little high. It is, to say the least, extravagant to describe as 'bestial and diabolical' the selfish manœuvres of the frightened and the hysterical. It is also patently unfair to the many thousands who met the Black Death with courage and charity. But though Niebuhr's words may seem fantastical he still had a valid and important point. Any history of the Black Death which ignored its impact on the minds of its victims would be notably incomplete. It was an impact whose effects endured. The resilience of mankind is perpetually astonishing and within only a few years the horrors of the plague had been thrust from the forefront of their minds. But no one can live through a catastrophe so devastating and so inexplicable without retaining for ever the scars of his experience.

It is a truism to say that, in the Middle Ages, a man's mental health and the public and private morality to which he deferred was inextricably involved with his relationship with the Church. His faith was unquestioning and his psychological dependence upon its institutions complete. Any blow suffered by the Church was a direct blow to his own morale. Any discussion of his state

of mind after the Black Death must start by considering how far
the condition of the Church had been modified by the events of the
preceding years. There can be little doubt that it had changed, and
changed almost exclusively for the worse.

Fairly or unfairly, medieval man felt that his Church had let him
down. The plague, it was taken for granted, was the work of God,
and the Church assured him, with uncomfortable regularity, that
he had brought it on his own head. 'Man's sensuality . . . now
fallen into deeper malice,' had provoked the Divine anger and he
was now suffering just retribution for his sins. But the Church
must have seen what had been going on over the previous years
and decades, yet had given no sign that the patience of the
Almighty was being tried too high. It would, perhaps, not have
been reasonable to have expected protection from the wrath of
God but surely it was not too much to ask that the Church, presumably better equipped than anybody else to predict a coming
storm, should have given some warning of the danger that mankind was courting? Instead, there had been no more than the
routine remonstrance which made up the repertoire of every
preacher. All that the Church had done was wait until it was too
late and then point out to their flock how wicked they had been.

The villagers observed with interest that the parish priest was
just as likely, indeed more likely, to die of the plague than his
parishioners. God's wrath seemed just as hot against Church as
against people: a significant commentary on those preachers who
denounced all their fellows with such tedious zest. So little hard
evidence survives that any generalisation about relationships
within society can be little more than guess work. Yet it would
have been surprising, in a community so credulous and so
deeply religious, if the village priest, the Man of God, had not
been accorded, not only the respect due to a figure of temporal
power, but also a tinge of awe appropriate to one who enjoyed a
special relationship with the Almighty. That the parson was mortal, everyone knew, that he ate, drank, defecated and in due course
died. Often, indeed, he came from the same village as his flock
and had relations living near him to testify that he was but flesh
and blood. Yet, with his ordination, surely he acquired too a
touch of the superhuman; remained a man but became a man
apart? After the plague, his vulnerability so strikingly exposed,
all trace of the superhuman must have vanished.

But he might at least have hoped that what he lost in awfulness he would regain in the sympathies of his flock. The priests, after all, suffered and died at the side of the laity; were always, indeed, among the most likely victims. Yet the slender evidence that exists shows that they lost in popularity as a result of the plague. They were deemed not to have risen to the level of their responsibilities, to have run away in fear or in search of gain, to have put their own skins first and the souls of their parishioners a bad second. It was not only captious critics like Langland or Chaucer who so accused them but their own kind. It was the Bishop of Bath and Wells who taxed them with lack of devotion to their duty;[1] a monk who wrote: 'In this plague many chaplains and hired parish priests would not serve without excessive pay';[2] the chronicler of the Archbishops who complained that: '. . . parishes remained altogether unserved and beneficed parsons had turned away from the care of their benefices for fear of death.'[3] Such criticism must have reflected a general view.

One of the most perplexing features of the Black Death is the reconciliation of this criticism with the outstandingly high mortality among the priests. We have already considered the factors which contributed to this high death rate. There is much which remains uncertain. But what seems clear beyond contradiction is that, if the parish priest had chosen to devote his superior wealth and the privileges conferred by his status solely to preserving his own person, he would have stood a better chance than his parishioners of survival. The fact that he suffered more proves that he cannot altogether have shirked his duty. The picture which one forms to explain this seeming ingratitude on the part of the people towards their priests is that of a clergy doing its daily work but with reluctance and some timidity; thereby incurring the worst of the danger but forfeiting the respect which it should have earned. Add to this a few notorious examples of priests deserting their flocks and of conspicuous courage on the part of certain wandering friars, and some idea can be formed of why the established Church emerged from the Black Death with such diminished credit. The contempt of contemporaries may not have been justified but it was still to cost the Church dear over the next decades.

There is little doubt that those who wished to criticise the Church found plentiful grounds on which to do so during the

next few years. On the whole, in plague as in war, those who take
most care of themselves live on while those who expose themselves perish. The best of the clergy died, the worst survived.
Obviously this is an over-simplification; many good men were
lucky as well as less noble who were unfortunate. But the abrupt
disappearance of nearly half the clergy, including a disproportionately great number of the brave and diligent, inevitably put
a heavy strain on the machinery of the Church and reduced
its capacity to deal effectively with movements of protest or
revolt.

It does not seem that the new recruits who took the place of
the dead were spiritually or, still more, educationally of the
calibre of their predecessors. During and immediately after the
plague the usual rules governing the ordination of priests were
virtually abandoned. Many men who found themselves widowed
took holy orders when already middle aged. In the diocese of
Bath and Wells a priest was admitted to holy orders even though
his wife was still alive and had not entered a cloister on the somewhat shaky grounds that 'she was an old woman and could remain in the world without giving rise to any suspicions.'[4] The
Bishop of Norwich obtained a dispensation to allow sixty clerks
aged twenty-one or less to hold rectories on the grounds, more or
less categorically stated, that they would be better than nothing.
In Winchester, in 1349 and 1350, twenty-seven new incumbents
became sub-deacons, deacons and priests in successive ordinations and thus arrived virtually unfledged in their new offices.[5]
The Archbishop of York was authorised to hold emergency
ordinations in his diocese.[6]

There is, of course, no reason to assume that a priest will be any
the worse for having been married and taken to his new vocation
late in life. Knighton it is true, referred to such recruits dis-
paragingly: '. . . a very great multitude of men whose wives had
died of the pestilence flocked to Holy Orders of whom many
were illiterate and no better than laics except in so far as they could
read though not understand';[7] but Knighton, as a Canon Regular,
had anyway little use for the secular priesthood. Nor did the fact
that some of the new priests were unusually young when ordained mean that they would lack a sense of vocation or fail to
become as good as their predecessors when they had gained experience. But, in the unseemly rush to fill the gaps, many un-

suitable candidates must have been appointed and many novices thrust unprepared into positions of responsibility. Certainly in the first few years after the plague, when society was slowly pulling itself together, the Church must have been singularly ill-equipped to give a lead.

The ill wind of the plague blew some good even to the clergy. Coulton has calculated that, before the Black Death, the majority of livings in lay presentation were given to men not yet in priests' Orders, often not even in Holy Orders at all.[8] Analysing the figures for four dioceses over a long period before 1348 he found that 73.8 % of the parishes were served by 'non-priests' unable to celebrate Mass, marry their parishioners or administer last rites. Most of these amateur rectors appointed professional curates to do their work but, though the souls of the parishioners might not be imperilled, the situation whereby prosperous absentees appointed qualified priests for the smallest possible wage to do the work which they were incompetent or unwilling to do themselves was not a happy one for Church or laity. During the Black Death the situation altered. The great majority of institutions went to ordained priests. The change survived the plague and an analysis of the period after the Black Death showed that the percentage of active priests in charge of benefices had more than doubled to 78 %.

But among these novice priests and the survivors of the plague there was noticeable a new acquisitiveness; a determination to share in the wealth which fell free for the taking after the Black Death. Such a determination was understandable. Many of their demands were anyway perfectly justified; the economic difficulties of certain parish priests have already been mentioned[9] and it was often literally impossible for the parson to live on what the plague had left him of his income. But insistence upon his financial due is rarely becoming to a minister of the Church and sometimes they were greedy and excessive in their exactions. 'A man could scarcely get a chaplain to undertake any church for less than £10 or ten marks,' grumbled Knighton. 'And whereas, while there had been plenty of priests before the plague and a man might have had a chaplain for five or six marks or for two marks and his daily bread, at this time there was scarce anyone who would accept a vicarage at £20 or twenty marks.'[10] Archbishop Islip had no doubt that 'the unbridled cupidity of the human race'

was at work among the priesthood. '. . . the priests who still survive, not considering that they are preserved by the Divine will
from the dangers of the late pestilence, not for their own sakes,
but to perform the Ministry committed to them for the people of
God and the public utility,' were neglecting their duties and seeking better paid conditions. So bad had things got that, unless the
priests at once mended their ways, 'many, and indeed most of the
churches, prebends and chapels of our and your diocese, and indeed of our whole Province, will remain absolutely without
priests.'[11]

This spectacle of priests haggling for extra pay and abandoning
their parishioners if better pickings were to be had elsewhere was
admirable material for those who were anyhow disposed to expect the worst of officers of the Church. 'Silver is sweet' commented Langland bitterly:

'Parsons and parish priests complained to the Bishop
That their parishes were poor since the pestilence time
And asked leave and licence in London to dwell
And sing requiems for stipends, for silver is sweet:'

while Chaucer put into the Reeve's mouth the mocking words:

'For Hooly chirches good moot been despended
On hooly chirches blood that is descended
Therfore he wolde his hooly blood honoure,
Though that he hooly churche sholde devoure.'*

Some writers have also ascribed to the Black Death the responsibility for an increase in the number of pluralities among
those who held benefices.[12] It would not have been surprising
if the dearth of priests had led to more cases in which a single
parson held two or more benefices but, though this seems to
have been the result in certain continental countries, in England
the 'great increase in the practice' to which Gasquet referred
did not take place. On the contrary the evidence points, if
anything, to the existence of less pluralities after the Black Death
than before it. Certainly the great increase in the number of or

---

* 'For Holy Church's goods should be expended
On Holy Church's blood, so well descended
And holy blood should have what's proper to it
Though Holy Church should be devoured to do it!'

dained priests appointed to benefices was likely to lead to a smaller proportion of non-resident parsons in the future.

The monasteries, on the whole, were still worse affected than the clergy. Including monks, nuns and friars the total population of the religious houses in England shortly before the Black Death had been something near 17,500.[13] Not far short of half these appear to have perished in the two years of the epidemic; probably more than half the friars and rather less than half the monks and nuns. In the seven monastic houses for which Snape has figures, the population dwindled by 51% between 1300 and the end of the plague [14] though some of this must be attributed to declining numbers between 1300 and 1348. Numbers were never to rise again to their earlier peak. Some houses, of course, suffered far worse than others; a few were virtually obliterated, a few left almost unscathed. Recovery too was fast in some places while in others it never took place at all. At Durham, Furness and Cleeve, numbers were so reduced that the refectory and dormitory had to be cut down in size proportionately. St Albans fell from a hundred to fifty monks and found even this figure difficult to maintain over the two centuries before the dissolution. Yet two of the greatest of England's religious leaders came to the fore at this period and new monastic colleges at Durham and Canterbury were founded shortly after it.

But the blow to the prestige and power of the monasteries did not stem only from their dwindling membership. The enormous number of chantries endowed in parish churches during and immediately after the Black Death inevitably detracted from the significance of the monasteries in the eyes of the people.[15] The high level of employment and new, exciting opportunities won away many of the more ambitious from spiritual pursuits; hitherto in the Middle Ages the Church, in one form or another, had offered to those who were not of noble birth almost the only prospect of economic or social advancement—now other possibilities were beginning to open. Many of the monks had grown accustomed to a way of life which was always comfortable and sometimes luxurious. With tithes unpaid and manorial incomes crippled, the always precarious economics of the worse-run monasteries slipped into deficit.[16] Debts quickly accumulated.

More than a hundred abbots succumbed to the plague. Not only did this mean a great loss in financial acumen and expertise but also in revenue, since the Crown took over the monastic income while its leadership was vacant and exacted a heavy fine before the new appointment could be made official.[17] The economic difficulties of the monasteries do not stem entirely from the Black Death; some houses were in trouble already while others managed to survive with their affluence unaffected. But the plague was certainly the most dramatic and probably the most important element in their decline.

To lose wealth and worldly power does not, of course, automatically imply a corresponding loss of spiritual grace. It is, indeed, more commonly argued that riches of the spirit accrue in inverse ratio to riches of the world. But there is little reason to believe that the new poverty of the monks brought with it any significant access of religious fervour—on the contrary, such evidence as there is indicates that the reverse was true. Wadding's denunciation of his own order, the Franciscans, is well-known:

'It was because of this misfortune [the Black Death] that the monastic Orders, in particular the mendicants, which up to this date had been flourishing, both in learning and in piety, now began to decline. Discipline became slack and faith weakened, both because of the loss of their most eminent members and the relaxation of rules which ensued as a result of these calamities. It was in vain to look to the young men who had been received without proper selection and training to bring about a reform since they thought more about filling up the empty houses than about restoring the lost sense of authority.'[18]

Though the mendicant orders were included in Wadding's strictures it seems, nevertheless, that they emerged from the plague years with heightened credit. Whether they were really more selfless and courageous than the parish clergy could hardly have been assessed by a contemporary, let alone to-day. The very fact that they had no territorial responsibility increased their chances of making an impression on the laity. When the parish priest performed his duty he was no more than a familiar figure doing what he had always done. The mendicant friar, descending as from heaven on a beleaguered village, was greeted with an enthusiasm which his better-established colleague rarely knew. But however this may have been, it does seem certain that their way of life precluded the display of materialism and even cupidity which was so marked among the priesthood. It was not only in England but on the mainland of Europe as well that the mendicants gained in authority and wakened the angry jealousy of their rivals.

In 1351, a counter-attack was launched. A petition, signed by a multitude of senior churchmen, was presented to Pope Clement VI, appealing for the abolition of the mendicant orders or, at least, that their members should be forbidden to preach or to hear confession. The Pope's reply at once defended the mendicants and provided a staggering indictment of the clergy. 'And if their preaching be stopped,' he asked, 'about what can you preach to the people? If on humility, you yourselves are the proudest of the world, arrogant and given to pomp. If on poverty, you are the most grasping and most covetous . . . If on chastity—but we will be silent on this, for God knoweth what each man does and how many of you satisfy your lusts.' He accused them of wasting their wealth 'on pimps and swindlers' and neglecting the ways of God.[19]

If there were any doubt that the Church in Europe was not generally admired or, indeed, deserving of admiration, in the years that followed the Black Death it would surely be settled by this astonishing attack delivered by a Pope on his own priests. Clement VI was himself by no means dedicated to austerity and was generally disinclined to rebuke too harshly the peccadilloes of the flesh. To have been provoked into such invective, he must have felt himself tried very far. That he was doing no more than voice the opinion of the people at large can not be doubted; that he himself, with his superior sources of information and personal responsibility for the doings of the Church, should have endorsed that opinion is a clear verdict of guilty against the priesthood.

Paradoxically, the decades that followed the plague saw not only a decline in the prestige and spiritual authority of the Church but also a growth of religious fervour. One example of this, to which we have already referred, was the large number of chantry chapels which were opened all over England. There was a spate of church building throughout Europe; the Cathedral of Milan is probably the most conspicuous example but the countrysides of England,

France and Italy are rich in village churches begun between 1350 and 1375. In Italy, nearly fifty new religious holidays were created; a move presumably inspired by relief at the end of the plague and fear lest, unless propitiated, the Almighty might once more unleash the whirlwind.[20] The number of pilgrims to Rome and other centres did not fall off even though a third of those who might have made the journey were now dead. In some cases, indeed, the number rose substantially after 1349 and 1350.[21]

In Florence, the Company of Or San Michele, a society with various religious and philanthropic functions, received donations worth 350,000 florins during or immediately after the Black Death. Most of this came in the form of legacies.[22] Though such generosity on the part of the rich could clearly not continue at a panic level once the plague was over, repeated threats that new disasters were imminent ensured that the flow continued at a healthy level.[23] It is easy to portray such charity as no more than the insurance policy of a rich man, well-informed about the dangers of moth and rust and prudently piling up treasure in heaven. So, indeed, it was. But, as Dr Meiss has demonstrated in his brilliant analysis of the impact of the Black Death on the Italian bourgeoisie, the confidence of the wealthy Florentine in the validity of his most fundamental assumptions had been badly shaken. The fortunes which they had amassed seemed more a source of guilt than of security. Their gifts to charity or to the Church were inspired partly by the urge to propitiate the angry god and partly by a shrewd calculation that some minor redistribution of wealth might avert discontent and civil disorders in the future. But they also reflected a strong distaste, a revulsion almost from the prosperity and luxury which they had long been accustomed to cherish as the most indispensable feature of their lives.

Yet, perhaps even more than this, the frenzied charity in which the rich of Europe indulged during and after the Black Death, demonstrated their faith in the one institution where it seemed a proper sense of social discipline survived. Discredited the Church might be in the eyes of many but, to the nobles and the monied élite, it was still the dyke which held back the flood of anarchic insurrection. Unless it were shored up then everything, it seemed, might be swept away. The rich gave eagerly so that the clergy might beautify their buildings and enhance their standing in the world. 'Their own position seriously threatened, they felt sustained by the assertion in art of the authority of the Church and the representation of a stable, enduring hierarchy.'[24]

The religious revival had therefore a strong element of the conservative. But this was only one strand in a complex which contained at least as much of the violently radical. The second half of the fourteenth century was marked in many countries by resentment at the wealth and complacency of the Church and fundamental questioning of its philosophy and its organisation. In England it was the age of Wyclif and of Lollardy, a new and aggressive anti-clericalism to some extent made use of by ambitious men who envied the riches and influence of the Church, but drawing its strength from the discontent and disillusionment of the people at large. In Italy it was the great period of the Fraticelli, dissident Franciscans who believed that poverty was of the essence of Christ and that a rich church must be a bad one. These rebels, having been denounced as heretics by Pope John XXII thirty years before, now declared the Pope himself a heretic and rejected all sacraments except their own. All over Europe the co-fraternities, the *confréries*, grew up, as it were, in the shadow of the great religious orders. In essence they were more movements of withdrawal than of protest; lay groups of simple idealists who sought refuge in their own society from a harsh and vicious world. But their very existence was an implied criticism of the system which they rejected and inevitably they began to evolve habits and strike attitudes inimical to the orthodox organisations from which they had evolved.

Once again, as so often in the history of the Black Death, one must remember that *post hoc* is not necessarily *propter hoc*. The second half of the fourteenth century was a time of spiritual unrest, of pertinent questioning of the values and of the conduct of the Church, of disrespect for established idols and a seeking for strange gods. Though the tempo of events would have been different, changes would have taken longer to bring about, resistance would have been more intense and reaction more immediate; in the long run things would have followed the same course, even though there had never been a plague. Dr Levett's judgement has already been quoted:

'The Black Death did not, in any strictly economic sense, cause the Peasants' Revolt or the breakdown of villeinage, but it gave birth, in many cases, to a smouldering feeling of discontent, an inarticulate desire for change . . .'

Coulton has suggested that the passage would read as well if 'theological' were substituted for 'economic' and 'Reformation' for 'Peasants' Revolt.'[25] The comment is a fair one. The Black Death did not cause the Reformation, it did not stimulate doubts about the doctrine of the Transubstantiation; but did it not cause a state of mind in which doctrines were more easily doubted and in which the Reformation was more immediately possible? Did it not break down certain barriers, psychological as well as physical, which otherwise might have impeded its advent? Wyclif was a child of the Black Death in the sense that he belonged to a generation which had suffered terribly and learned through its sufferings to doubt the premises on which its society was based. The Church which he attacked was a victim of the Black Death because of the legion of its most competent and dedicated officers who had perished and, still more, because of the honour and respect which it had forfeited in the minds of men. The Church continued as an immensely potent force in the second half of the fourteenth century but the unquestioned authority which it had been used to exercise over its members was never to be recovered. To this decay the Black Death made a signal contribution.

Matteo Villani, the Florentine historian, devoted a passage to the effects of the Black Death on those who were fortunate enough to survive it:[26]

'Those few discreet folk who remained alive,' he wrote, 'expected many things, all of which, by reason of the corruption of sin, failed among mankind, whose minds followed marvellously in the contrary direction. They believed that those whom God's grace had saved from death, having beheld the destruction of their neighbours . . . would become better-conditioned, humble, virtuous and Catholic; that they would guard themselves from iniquity and sin and would be full of love and charity towards one another. But no sooner had the plague ceased than we saw the contrary; for since men were few and since, by hereditary succession, they abounded in earthly goods, they forgot the past as though it had never been, and gave themselves up to a more shameful and disordered life than they had led before. For, mouldering in ease, they dissolutely abandoned themselves to the sin of gluttony, with feasts and taverns and delight of delicate viands; and again to games of hazard and to unbridled lechery, inventing strange and unaccustomed fashions and indecent manners in their garments . . .'

'. . . Men thought that, by reason of the fewness of mankind, there should be abundance of all produce of the land; yet, on the contrary, by reason of men's ingratitude, everything came to unwonted scarcity and remained long thus; nay, in certain countries . . . there were grievous and unwonted famines. Again, men dreamed of wealth and abundance in garments . . . yet, in fact, things turned out widely different, for most commodities were more costly, by twice or more, than before the plague. And the price of labour and the work of all trades and crafts, rose in disorderly fashion beyond the double. Lawsuits and disputes and quarrels and riots rose elsewhere among citizens in every land . . .'

Contemporary chronicles abound in accusations that the years which followed the Black Death were stamped with decadence and rich in every kind of vice. The crime rate soared; blasphemy and sacrilege was a commonplace; the rules of sexual morality were flouted; the pursuit of money became the be-all and end-all of people's lives. The fashions in dress seemed to symbolise all that was most depraved about the generation which survived the plague. Who could doubt that humanity was slipping towards perdition when women appeared in public wearing artificial hair and low-necked blouses and with their breasts laced so high 'that a candlestick could actually be put on them.' When Langland dated so many of the vices of the age 'sith the pestilens tyme' he was speaking with the voice of every moraliser of his generation.

No doubt the rodomontades of the virtuous were often over-stated and the plague blamed for much for which it was not responsible. Hoeniger for one has suggested that 'the low state of morals belonged to the period and was no worse after the epidemic than before.'[27] But this cannot be the whole answer, nor was it only the impressionable contemporary chronicler who has recorded the phenomenon. In her study of Orvieto, Dr Carpentier has found ample evidence that the Black Death was followed by an immediate and sharp decline in public morality. There were many more cases of maltreatment of orphans, more people carried arms, the strict rules governing female dress were relaxed and there was a considerable increase in the number of prosecutions and convictions for every kind of crime.[28]

Such self-indulgence strikes one to-day as a curiously illogical reaction to the disaster which had been so painfully survived. Medieval man in 1350 and 1351 believed without question that the Black Death was God's punishment for his wickedness. This time he had been spared but he could hardly hope for such indulgence to be renewed if his contumacious failure to mend his ways stung God into a second onslaught. The situation, with sin provoking plague and plague generating yet more sin, seemed to have all the makings of a uniquely vicious circle, a circle from which he could only hope to escape by a drastic mending of his ways. Yet, undeterred, he continued on his wicked course against a background of apocalyptic mutterings prophesying every kind of doom.

In spite of the diatribes of Matteo Villani and the more prosaic statistics of Dr Carpentier it is difficult to take altogether seriously the sins of the post-plague generation. Laxness there certainly was, but it was the laxness which comes through relief from almost intolerable tension and the enjoyment of more money than one has been used to spending. In a stimulating essay, Mr J. W. Thompson has drawn an interesting if sometimes strained analogy between the reactions of the population after the Black Death and after the Great War of 1914-18.[29] In both cases he finds the same complaints about the immorality and instability of those who survived. The gloomy relish with which their conduct was denounced can be matched in the Naples of 1944, the Paris of 1815 or, indeed, in almost any situation where human beings recuperate from some extensive disaster. The decline in morality should not be ignored but nor should it be imagined that the Europeans who survived the Black Death had any very special attributes in the way of wickedness.

Nevertheless, no society could endure the punishment which the plague had meted out and emerge without serious strains. One sign of this, already mentioned, was the damage done to the relationship between priest and laic; another, the tension that arose between different groups within society, in particular between rich and poor. Renouard has examined the result of this in France.[30] In the country, he says, the landowner was generally impoverished while the lot of the peasant, on the whole, improved. An up-and-coming peasantry clashed with an impoverished ruling class which sought to regain its former prosperity at the expense of its tenants. In the cities the situation was very different. Here the laws of inheritance ensured that those who survived among the rich accumulated still greater fortunes while the poor, with nothing to inherit, were economically no better off. A newly rich bourgeoisie joyfully oppressed a defenceless proletariat. In the countryside the gap between rich and poor narrowed, in the cities it widened; in both cases relations deteriorated as a result. The second half of the fourteenth century in France was peculiarly rich in social disorder and the scars left by the Black Death were at least in part responsible for the rural insurrection of Jacques in 1358 and of Tuchins in 1381, and for the risings of the weavers in Ghent in 1379, of the Harelle at Rouen in 1380 and of the Maillotins in Paris in 1382.

The phenomenon recorded by Renouard seems to have been by no means invariable. In Albi, to take one example, little difference was perceptible in the social structure of the city as a result of the plague. Almost everybody was richer than before but, on the whole, the wealth of the dead seems to have been shared out among the living without any further distortion of what was already a formidably inequitable pattern.[31] But even where no extra economic motive arose, the Black Death left a legacy of mistrust between classes. No one could protect themselves against infection but the rich were at least able to take to flight. The bishops, the territorial magnates, the more affluent merchants, took refuge on their country manors and left the city to look after itself as best it could. It was not to be expected that they would meet with much enthusiasm on their return. It was as if Mayfair or the Sixteenth *Arrondissement* had emptied themselves in time of war and the inhabitants returned when the danger was passed, expecting a welcome from those whom they had deserted.

The impression that the rich escaped the worst of the plague was largely illusory. Many stayed in the cities and, of those who fled, many also found that their rural fastnesses offered no protection. But they did suffer less and their luck was obvious to the less fortunate. An isolated statistical illustration comes from Teruel in Aragon.[32] In 1342, 33.7% of those citizens liable to pay tax did not do so because they had so little money that they were exempted. By 1385, the proportion had dropped to 10.4%. One can accept that, immediately after the Black Death, some redistribution of wealth could have accounted for a drop in the numbers of the poor. But so sharp a fall in their numbers, continuing thirty-five years later, suggests strongly that the poor in the city must almost have been wiped out in the Black Death or subsequent epidemics. The victims did little to express their resentment—there was, indeed, very little that they could do—but a new and potentially dangerous element of class hatred had sprung up.

Professor Russell has suggested that, though the numbers of the peasants may have dropped, at least those who were left were likely to be more healthy. Even if the rule of the survival of the fittest did not apply, the extra food available after the mortality would ensure that they would fare better in future. So far as the Italian peasant, at least, was concerned, Professor Russell's rosy vision was quickly disallowed. According to Miss Thrupp, this unfortunate species suffered, in the fourteenth century, from acute over-exposure and protein deficiency leading to asthma, quinsy, erysipelas, various digestive and intestinal complaints and bad teeth.[33] Such complaints could only have been cured by a radical change of diet and of living conditions: the only possible benefit they might have gained from the plague was a marginally larger intake of anyhow monotonous and insalubrious food. The medieval peasant, it is clear, had little in the way of material gain to set against the anguish which he had suffered.

But if one were called on to identify the hall-mark of the years which followed the Black Death, it would be that of a neurotic and all-pervading gloom. 'Seldom in the course of the Middle Ages has so much been written concerning the *miseria* of human beings and human life,' wrote Hans Baron, going on to refer to '. . . the pessimism and renunciation of life which took possession of mankind in the period following the terrible epidemics in the middle of the fourteenth century.'[34] It was a gloom which fed upon extreme uncertainty and apprehension. The European of this period lived in constant anticipation of disaster. The apparition of Antichrist was announced many times and in many places. Floods, famines, fire from heaven were perpetually around the corner. The Turks and Saracens planned a descent on Italy; the Engish on France; the Scots on England. Medieval man, in sober fact, had more than enough to worry about. Now his imagination ran riot.

Perhaps the factor which contributed most towards his demoralisation was his almost total ignorance of the workings of his world. Severe though the limitations may be on modern man's ability to control his destiny he now has a rudimentary understanding of the way in which the forces which dominate him achieve their irresistible effect. Once a danger is understood then half its terrors are gone. From the tiny patch of fitful light which played within the circle of their comprehension our fore-fathers stared aghast into the darkness. Strange shapes were moving, but what they were they did not know and hardly dared to speculate; strange sounds were heard but who could say from where they came? Everything was mysterious, everything potentially dangerous; to stay still might be perilous, to move fatal. The debauchery and intemperance of which we have spoken was the protective device of frightened men who drank to keep their spirits up, who whistled in the dark. And yet their frenetic gaiety only served to accentuate the gloom that lay underneath. 'Eat, drink and be merry, for to-morrow we die'; but to-morrow seemed very close and the food and drink could never subdue for long the fear of death.

'Psychologists and sociologists,' wrote Mollaret,[35] 'know that man reacts to violent pain by flight, by violence or by sublimation. The plague stirred up these three reactions. Flight took the form of a stampede towards altars and processions; doctors and quacks; workers of miracles and visionaries. Violence found its outlet in the massacre of the Jews or those believed to have spread the plague, in the hysteria of the Flagellants, often in suicide. Sublimation was the works of the artists . . .'

It is above all in the works of the artist that the mood of the age finds its most vivid expression. The favourite themes were those of suffering and of retribution; Christ's passion or the tortures of hell. Orcagna's great, though now sadly battered fresco, the 'Triumph of Death' in the church of Santa Croce in Florence, painted immediately after the Black Death, sets the pattern for the school.[36] In this lugubrious composition a King and Queen are hunting with their suite. They turn a corner and see three open graves before them. Each one contains a corpse; all worm-eaten, one blackened, one covered with snakes, one with belly distended, all wearing crowns. Golden lads and girls all must, as chimney sweepers come to dust. The King, it is clear from his expression, has not failed to draw the proper moral. To the left a party of gay debauchees are holding an alfresco feast and giving every sign of thoroughly enjoying themselves. They do not notice Death, a clawed harpy, preparing to swoop upon them. To the right the lepers and the blind, the halt and the lame plead to be relieved of their sufferings.* Death ignores them. Another moral: Death prefers to pick his victims from those who wish to live. Unsubtle, didactic, but terrifying in its force and in its revelation of the fear and hopeless pessimism that occupied the mind of the painter and of his public.

It was characteristic of the age that Christ should often be portrayed as an angry and minatory figure; that Death should be personified in a higher proportion of paintings than before or after, that the cult of St Sebastian should become fashionable, that the story of Job should appear almost for the first time in Tuscan panels and frescoes. In pictures of the Virgin dating from before the plague she is often seen protecting monks and nuns from the wrath of God; significantly, after the plague, her mantle is extended to cover all Christian beings as well.[37] Mankind, it is clear, could do with all the protection it could get.

It would be wrong to suggest that the Black Death was solely and directly responsible for the metamorphosis. 'In the 13th century,' wrote Émile Mâle,[38] 'all the inspiring aspects of Christianity are reflected in art—kindness, humanity, love . . . In the 15th century this light from heaven has long ceased to shine. Most of the works from this period . . . are sombre and tragic. Art offers only a representation of grief and of death.' Yet Mâle saw this evolution at its most rapid in the last quarter of the fourteenth century, the fruit not of the plague itself but of the great wave of terror and dismay which engulfed Europe even after the plague was passed. Whatever the exact timing, however, the mood of the age pervades its religious art and the explanation of that mood forces itself upon the historian.

In a relatively primitive society religion and magic are seldom far apart; the shadow of the sorcerer lurks in the churchyard, the relics of the saints and the bones of the witch doctor's mumbo-

* Reproduced as the frontispiece to this book.

jumbo provide alike a picturesque pendant to an inner mystery. The terrors of the Black Death drove man to seek a more intense, a more personal relationship with the God who thus scourged him, it led him out of the formal paths of establishment religion and, by only a short remove, tumbled him into the darkest pit of Satanism. The Europeans of the 1350s and 1360s were no more saints or devils than their ancestors but such emotional disturbance had been generated that they were often within a step of believing themselves one or the other. They had been tested to the uttermost and even a touch was henceforth enough to tip them from their precarious balance. For anyone who had lived through the Black Death, hysteria could never be far away.

We have already referred to Mr Thompson's analogy between the after-effects of the Black Death and the Great War of 1914-18.[39] In both cases, he says, complaints of contemporaries were the same: 'economic chaos, social unrest, high prices, profiteering, depravation of morals, lack of production, industrial indolence, frenetic gaiety, wild expenditure, luxury, debauchery, social and religious hysteria, greed, avarice, maladministration, decay of manners.' In both, the immense loss of life and 'psycho-physical shock' made it a long time before the vitality and the initiative of the survivors was regained. In both the 'texture of society' was modified; new openings were created; the old nobility largely passed away and parvenu upstarts took their place; chivalry and courtesy vanished, manners became uncouth and brutal, refinement in dress disappeared. In both the administrative machine and the Church were almost crippled. Thousands of ignorant, incompetent, dishonest men were thrust abruptly into positions of authority far beyond their merit. In both, in a word, the whole population was 'shell-shocked,' a state from which they were not fully to emerge for many years.

There is, of course, much which is exaggerated or unacceptable in this thesis. Neither after 1918 nor after 1350 did the 'old nobility' pass away; chivalry and courtesy did not vanish; dress and manners evolved, perhaps more dramatically than usual, but certainly not with the drastic absoluteness suggested by Thompson. But the analogy is still of value in that it conveys an impression, expressed in contemporary terms, of the magnitude of the experience in which medieval man had been involved. The two experiences are properly comparable but comparison can only show how much more devastating the Black Death was for its victims than the Great War for their descendants. Their chances of death were of course immeasurably higher; even the front-line infantry man had a better chance of surviving the war than the medieval peasant the plague. Another distinction perhaps even more important for the morale of those involved, was the omnipresence of the Black Death. The First World War was more or less confined to contending armies on fixed battle lines; the second cast its net more widely but still left great areas virtually untouched. The Black Death was everywhere; in every hamlet and in every home. No escape was possible.

And then, in the Great War, the warring nations knew their enemies and knew, too, that they were merely mortal. They had a defined target to hate and to contend with. The plague victim could hate only his God or himself, with occasional not wholly convincing forays in persecution of Jews, lepers or other even less substantial surrogates. For the rest his affliction was totally mysterious and far the more dreadful for being so. And finally the Great War was spread over more than four years; for each locality the great pestilence worked its mischief in a tenth the time. It can be argued that protracted agony is worse than a sudden and explosive shock, but if one is considering the impact on the mind of the survivor then it is surely the second which will produce the graver consequences.

The 'shell-shock' which Mr Thompson finds in the survivors of the two catastrophes should therefore have been more violent and more lasting among those who endured the Black Death. But is this not particularly striking generalisation of any value? What form did the 'shell-shock' take? Was it the kind of shock which galvanised into action or which stunned into apathy? Is it possible to detect any significant and consistent change in the attitude of medieval man between the first and second halves of the fourteenth century? Does the statement that modern man was forged in the crucible of the Black Death have any real validity? Does it, indeed, mean anything at all?

To some at least of these questions this book may provide a tentative answer. But it is only necessary to pose them to see how far we are, and will always be from a definitive solution. In a field so amorphous any attempt at the precise or the categoric would be futile. But if one were to seek to establish one generalisation, one cliché perhaps, to catch the mood of the Europeans in the second half of the fourteenth century, it would be that they were enduring a crisis of faith. Assumptions which had been taken for granted for centuries were now in question, the very framework of men's reasoning seemed to be breaking up. And though the Black Death was far from being the only cause, the anguish and disruption which it had inflicted made the greatest single contribution to the disintegration of an age.

'Faith disappeared, or was transformed; men became at once sceptical and intolerant. It is not at all the modern, serenely cold, and imperturbable scepticism; it is a violent movement of the whole nature which feels itself impelled to burn what it adores; but the man is uncertain in his doubt, and his burst of laughter stuns him; he has passed as it were, through an orgy, and when the white light of the morning comes he will have an attack of despair, profound anguish with tears and perhaps a vow of pilgrimage and a conspicuous conversion.'[40]

Jusserand's classic description of the European in the second half of the fourteenth century captures admirably the twin elements of scepticism and timorous uncertainty. The generation that survived the plague could not believe but did not dare deny. It groped myopically towards the future, with one nervous eye always peering over its shoulder towards the past. Medieval man during the Black Death, had seemed as if silhouetted against a background of Wagnerian tempest. All around him loomed inchoate shapes redolent with menace. Thunder crashed, lightning blazed, hail cascaded; evil forces were at work, bent on his destruction. He was no Siegfried, no Brunnhilde heroically to defy the elements. Rather, it was as if he had wandered in from another play: an Edgar crying plaintively, 'Poor Tom's a-cold; poor Tom's a-cold!' and seeking what shelter he could against the elements.

Poor Tom survived, but he was never to be quite the same again.

Notes

CHAPTER ONE

p13 1 J. F. C. Hecker, *The Epidemics of the Middle Ages*, (Trad. B. G. Babington) London 1859, p11. cf. Abbé des Guignes. *Histoire des Huns*, Paris 1757.

14 2 *Voyages de Ibn-Batoutah*, Société Asiatique, Paris 1853.

3 De Smet. 'Breve Chronicon clerici anonymi.' *Recueil des Chroniques de Flandres*. Vol III p14.

4 *Storie Pistoresi*. Muratori 11, V p237.

5 *Chronicon Estense*. Muratori 15, III p160.

15 6 Hecker. op. cit. p21.

7 Haeser. *Archiv für die gesammte Medizin*. Jena, Vol II p29.

8 A. G. Tononi. 'La Peste dell'Anno 1348.' *Giornale Ligustico*. Genoa, Vol X 1883 p139.

9 G. Vernadsky. *The Mongols and Russia*. Yale 1959.

10 Haeser. op. cit. Vol II pp48-49.

16 11 C. S. Bartsocas. 'Two 14th Century Greek Descriptions of the "Black Death".' *Journ. Hist. Med.* Vol XXI No 4 1966, p395.

12 *Chronicon Estense*. op. cit. p160.

17 13 De Smet. op. cit. Vol III pp14-15.

14 S. d'Irsay. 'Notes to the origin of the expression: Atra Mors.' *Isis*, Vol 8 1926, p328.

15 R. Kjennerud. *Journal of the History of Medicine*. Vol III 1948, p359.

18 16 Gasquet. *The Black Death*. London 1908, p8.

17 J. Michon. *Documents inédits sur la Grande Peste de 1348*. Paris 1860, p11.

18 J. Nohl. *Der Schwarze Tod*. Potsdam 1924, p11.

19 Boccaccio. *Decameron*. Trans. J. M. Rigg. London 1930, p5.

19 20 e.g. Lechner. *Das grosse Sterben*. p15; or plague tractates of Ibn Khātimah or Gentile da Foligno.

21 *La Grande Chirurgie*. ed. E. Nicaise. Paris 1890, p171.

22 *Chronicon Galfridi le Baker*. ed. E. Maunde Thompson. Oxford 1889, p99.

20 23 Simon of Covino. *Bibl. de l'École des Chartes*. 1840-41 Ser. I Vol 2 p241.

20 24 J. P. Papon. *De la Peste ou Époques mémorables de ce Fléau*. Vol I p115.

21 25 Ibn Khātimah. Sudhoff. *Archiv. für Geschichte der Medizin*. Vol XIX 1927, p30.

26 Ibn al Khatib. *Sitzungsberichte der Königl: bayer*. Munich 1863 II p1.

27 Alfonso de Cordova. Sudhoff. *Archiv*. III p225.

28 'Utrum mortalitas . . .' Sudhoff. *Archiv*. XI p44. cf. Konrade of Megenberg. *Buch der Natur*. Ed. Pfeiffer. Berlin 1870.

22 29 Hirst. *The Conquest of Plague*. Oxford 1953, p28.

30 De Smet. op. cit. *Breve Chronicon*. Vol III p15.

23 31 'Tractatus de epidemia.' Michon. *Documents inédits*. op. cit. p46.

24 32 C. Creighton. *History of Epidemics in Britain*. Cambridge 1891, p175.

33 The most useful studies of bubonic plague which make special reference to the Black Death are Greenwood's *Epidemics and Crowd Diseases*, Pollitzer's *Plague* and Hirst's *The Conquest of Plague*.

25 34 J. Stewart. *The Nestorian Missionary Enterprise*. Edinburgh 1928, p209.

35 R. Pollitzer. *Plague*. W.H.O. Publication. Geneva 1954, p13.

26 36 Jorge. *Bull. Off. Int. Hyg. Publ*. Vol 25 1933, p425.

37 MacArthur. 'Old Time Plague in Britain.' *Trans. Roy. Soc. Trop. Med. Hyg*. Vol XIX p355.

27 38 *Reports on Plague Investigations in India*. No 39 of 1910.

39 Hirst. op. cit. p324.

40 Nohl. op. cit. p31.

28 41 Greenwood. *Epidemics and Crowd Diseases*. London 1935, p308.

42 op. cit. p28.

43 Greenwood. op. cit. p291.

CHAPTER TWO

30 1 *Documents inédits*. op. cit. p21.

2 See, in particular, A. R. Bridbury, *Economic Growth*. London 1962 cf. E. Miller, 'The English Economy in the 13th Century.' *Past and Present*. 1964, No 28 p21.

3 H. Nabholtz. *Camb. Econ. Hist. Eur*. Vol I 1941, p493.

31 4 E. Power. *Camb. Med. Hist*. Vol VII 1932, p731.

5 L. Genicot. *Camb. Econ. Hist*. Vol I, 2nd Edition, 1966, pp668-69.

6 M. Postan. *Camb. Econ. Hist*. Vol II 1952, p160.

32 7 L. Genicot. op. cit. p666.

8 G. Utterström. 'Climate Fluctuations and Population Problems in Early Modern History.' *Scan. Econ. Hist. Rev.* III 1955, pp3-47.

9 M. Postan. *Camb. Econ. Hist.* Vol I, 2nd Edition, 1966, p565.

33 10 H. S. Lucas. 'The Great European Famine of 1315, 1316 and 1317.' *Speculum.* Vol 5 1930, p355.

11 H. Pirenne. *Economic and Social History of Mediaeval Europe.* 1936, p193.

12 L. Genicot. op. cit. p673.

13 Ibid. p666. M. Postan. 'Some Economic Evidence of Declining Population in the Later Middle Ages.' *Econ. Hist. Rev.* 2nd Ser. Vol II 1950, p221.

34 14 *L'économie rurale et la vie des campagnes dans l'Occident médiéval.* Paris 1962.

15 *Econ. Hist. Rev.* 2nd Ser. Vol XVI 1963, p197.

16 B. H. Slicher van Bath. *Agrarian History of Western Europe.* New York 1963, p84.

17 e.g. R. Delatouche. 'Agriculture médiévale et population.' *Études Sociales.* 1955, pp13-23.

18 E. Carpentier. 'Autour de la Peste Noire.' *Annales, E.S.C.* Vol XVII 1962, p1092.

35 19 ed. Pfeiffer. Berlin 1870.

36 20 de Mussis. op. cit. p50.

21 *Chronicon Henrici Knighton.* R.S. 92 ii pp57-58.

22 *Piers Plowman.* Version B v.13.

37 23 C. Singer. *Proc. Roy. Soc. Med. (Hist. Med.)* Vol X 1917, p107.

38 24 Reprinted by Michon (p32), but the fullest text is that of H. E. Rebouis *Étude historique et critique sur la peste.* Paris 1888.

25 Sudhoff. *Archiv.* V p83.

CHAPTER THREE

40 1 Michael of Piazza (Platiensis). *Bibliotheca scriptorum qui res in Sicilia gestas retulere.* Vol 1 p562.

43 2 'La Peste Noire.' *Revue de Paris.* March 1950, p108.

3 André Siegfried. *Itinéraires des contagions: épidemies et idéologies.* Paris 1960, p114.

4 Coulton. *Black Death.* op. cit. p9.

5 *Monumenta Pisana.* Muratori 15. (1729 edition) p1021.

6 Sismondi. *Histoire des Républiques Italiennes du Moyen Age.* Paris 1826, Vol VI p11 et seq.

44   7   *Storie Pistoresi.* Muratori 11, V p224.

       8   Carpentier. *Une Ville devant la Peste.* op. cit. pp79-81.

       9   Sismondi. op. cit. p13.

   10   Giovanni Villani. *Cronica.* Florence 1845, Book 12 p92.

45 11   *Cambridge Mediaeval History.* Vol VII pp49-77.

   12   *Epistolae Familiares.* lib. VIII pp290-303.

46 13   Defoe's account of the Plague of London is an obvious rival but, since he was only seven years old in 1665, the term 'eye-witness' is perhaps loosely employed. The translation is that of J. M. Rigg in the Everyman edition (London 1930).

51 14   e.g. *Cronica Fiorentina.* Muratori 30, I p231.

   15   Giovanni Villani. *Cronica.* op. cit. Book II p122.

   16   E. Fiumi. 'La demografia fiorentina nelle pagine di Giovanni Villani.' *Archivio Storico Italiano.* 1950 Vol I p80.

52 17   E. Fiumi. *La popolazione . . . volterrano sangimignanese.* p280.

   18   W. M. Bowsky. 'The Impact of the Black Death upon Sienese Government and Society.' *Speculum.* Vol XXXIX No 1 1964, p18. Carpentier, op. cit. p135.

   19   *Black Death.* p28.

   20   Nohl. op. cit. pp6 and 26.

   21   *Chronicon Estense.* Muratori 15, III p162.

   22   *Cronica Gestorum ac factorum memorabilium civitatis Bononie.* Muratori 28, II p43.

   23   e.g. A. Doren. *Storia Economica dell' Italia nel Medio Evo.* Padua 1937, p579.

53 24   *Chronicon Estense.* op. cit. p162.

   25   Lorenzo de Monaci. *Chronicon de rebus Venetorum.* Brunetti. 'Venezia durante la peste.' *Ateneo Veneto.* 32, 1909.

   26   d'Irsay. 'Defence reactions during the Black Death.' *Annals of Medical History.* IX 1927, p171.

54 27   d'Irsay. op. cit. p174.

   28   Hecker. op. cit. pp58-59.

55 29   Alberto Chiappelli. 'Gli ordinamenti sanitari del Comune de Pistoia contra la peste de 1348.' *Arch. stor. ital.* Ser. IV Vol XX pp3-24. Anna Campbell. *The Black Death and Men of Learning.* p115.

56 30   *Une Ville devant la Peste. Orvieto et la Peste Noire de 1348.*

58 31   *Cronica Senese di Agnolo di Tura del Grasso.* Muratori 15, VI p555.

60 32   W. Bowsky. *Speculum.* Vol XXXIX op. cit. p34.

   33   Matteo Villani. Cronica. Florence 1846, Book 1 pp67-68.

61 34   S. M. Gromberger. 'St. Bridget of Sweden.' *American Catholic Quarterly Review.* Vol XLII 1917, p97.

35 D. Herlihy. 'Population, Plague and Social Change in Rural Pistoia.' *Econ. Hist. Rev.* 2nd Ser. Vol XVIII No 1 1965, p225.

36 A. Doren, op. cit. p579.

CHAPTER FOUR

63   1 De Smet. Vol II *Breve Chronicon.* p15.

    2 M. E. Lot. 'L'état des paroisses et des feux de 1328.' *Bibliothèque de L'École de Chartes.* Tome XC 1929.

    3 Y. Renouard. *Population.* Tome III 1948.

    4 J. R. Strayer. 'Economic Conditions in the Country of Beaumont-le-Roger.' *Speculum* XXVI 1951, p282.

    5 For the best resumé of the debate see E. Carpentier and J. Glénisson. 'La Dèmographie française en XIVe. Siècle.' *Annales E.S.C.* Tome XVII 1962 No 1 p109.

64   6 T. Wright. *Political Poems and Songs relating to English History.* p169.

    7 C. Anglada. *Étude sur les Maladies Éteintes.* p432.

65   8 R. Emery. 'The Black Death of 1348 in Perpignan.' *Speculum.* Vol XLII 1967 No 4 p611.

    9 cit. Crawfurd. *Plague and Pestilence in Literature and Art.* pp115-16.

66 10 De Smet. Vol 11 *Breve Chronicon.* pp16-17.

  11 *Storie Pistoresi.* Muratori 11, v p235.

  12 *Die Geschichte der Pest.* Giessen 1908, p57.

  13 Y. Renouard. 'La Peste Noire.' *Revue de Paris,* March 1950, p111.

  14 Knighton. *Chronicon.* R.S. 92 II p59.

  15 See, e.g. Lea. *History of the Inquisition.* Vol I p290.

67 16 De Smet. op. cit. Vol II p17.

  17 op. cit. p38 n 24 above.

70 18 C. Singer. *Short History of Medicine.* p69.

  19 C. Singer. 'Review of the Medical Literature of the Dark Ages.' *Proc. Roy. Soc. Med.* Vol X 1917, p107.

  20 *Geschichte der Chirurgie.* Berlin 1878, Vol 1 p673.

71 21 *La Grande Chirurgie.* op. cit. p171.

  22 *Archiv für Geschichte der Medizin.* 1910 onwards.

72 23 New York 1931.

  24 *Bibl. de l'École des Chartes.* (1840-41) Sér. 1 Vol 2 p240.

  25 Sudhoff. XIX p49.

73 26 *Primo de Epydimia.* Sudhoff. V p43.

  27 D. J. Colle. *De Pestilentia.* Pisa 1617, p570.

74 28 Compendium de Epydimia. op. cit. p60.

  29 cit. Campbell. p71. [Not in Sudhoff].

74	30	Sudhoff. XIX pp76-77.
	31	d'Irsay. *Annals of Medical History*. IX 1927, p174.
	32	*Gentile da Foligno*. Sudhoff. V p83.
75	33	ed. D. W. Singer. *Proc. Roy. Soc. Med. (Hist. Med.)* Vol 9 1916, p159.
76	34	Ibid.
78	35	Siméon Luce. *Bertrand de Guescelin*. pp69-73.
	36	cit. H. Martin, *Histoire de France*. Vol V p111.
	37	E. Carpentier and J. Glénisson. op. cit. p109.
	38	M. Mollat. 'La Mortalité à Paris.' *Moyen Age*. Vol 69 1963, p505.
	39	'Continuatio Chronici Guillelmi de Nangiaco.' *Soc. de L'Histoire de France*. Vol II 1844, pp211-17.
	40	p163 below.
79	41	*Black Death*. p55, basing himself on Géraud.
80	42	L. Porquet. *La Peste en Normandie*. Vire 1898, p77.
	43	Thierry. *Recueil des Monuments inédits de L'Histoire du Tiers État*. Vol 1 p544.
	44	'Chronicon majus Aegidii li Muisis.' De Smet. *Receuil des Chroniques de Flandres*. Vol 11 p280.
81	45	Y. Renouard. *Population*. Vol III 1948, p459.
	46	E. Carpentier. 'Autour de la Peste Noire.' op. cit. p1065.
83	47	*De la Peste* . . . op. cit. Vol I p123.

CHAPTER FIVE

84	1	Lechner. *Das Grosse Sterben in Deutschland*. Innsbruck 1884, p26.
	2	'Continuatio Novimontensis.' *Mon. Germ*. IX p675.
	3	G. Rath. *CIBA Symposium*. III 1956, p195.
	4	G. Sticker. *Die Geschichte der Pest*. Giessen 1908, p68.
	5	'Kalendarium Zwetlense.' *Mon. Germ*. IX p692.
	6	'Continuatio Novimontensis.' op. cit. p675.
85	7	Crawfurd. *Plague and Pestilence*. op. cit. p125.
	8	L. Porquet. *La Peste en Normandie*. Vire 1898, pp18-19.
	9	*Hierarchia catholica*. Vol I Münster 1913, cit. Campbell, p134.
	10	*Historia Suevorum*. Bk II pp309-10.
86	11	H. Reincke. 'Bevölkerungsverluste der Hansestädte durch den Schwarzen Tod.' *Hansische Geschichtsblätter*. Vol 72 1954, p88.
	12	F. Graus. *Histoire des paysans en Bohême*. Prague 1957; cit. Carpentier, 'Autour de la Peste Noire,' p1089.
	13	The best recent account of the Flagellant movement is that of G. Leff. *Heresy in the Later Middle Ages*. Vol II Chap VI Manchester 1967.

87 14 J. McCabe. *The History of Flagellation*. Girard, Kansas 1946.

88 15 Lea. *History of the Inquisition*. Vol II pp382-83.

16 J. Nohl. *Schwarze Tod*. op. cit. p303.

17 See, in particular, Matthew of Neueburg (Matthiae Neuewenburgensis). *Fontes Rerum Germanicarum*. ed. Boehmer. Stuttgart Vol IV 1868, pp266-67.

90 18 Henry of Herford. *Liber de rebus memorabioribus* ed. Potthast. Göttingen 1859, p281.

91 19 The translation is Babington's from Hecker's *Black Death*, p65.

20 Certain authorities prefer thirty-three and a half days.

93 21 *Mon. Germ*. NS. III p280.

94 22 R.S. 93 pp407-8.

23 *Historia Anglicana*. R.S. I p275.

95 24 R. Hoeniger. *Der Schwarze Tod in Deutschland*. Berlin 1882, p14.

96 25 Henry of Herford. op. cit. p282.

97 26 A. Lopez de Meneses. 'Documentos acerca de la Peste Negra en los dominion de la Corona de Aragon.' *Consejo Superior de Investigaciones Científicas*. Vol VI 1956, p301.

27 G. Sticker. *Die Geschichte der Pest*. op. cit. p59.

98 28 Lea. op. cit. Vol II p380.

29 H. Dubled. 'Aspects économiques de la vie de Strasbourg aux XIIIe et XIVe siècles.' *Archives de l'Église d'Alsace*, N.S. Tome VI 1955, pp23-56.

30 N. Cohn. *Pursuit of the Millenium*. London 1957, p124.

99 31 Ibid. p387.

100 32 Ilza Veith. 'Plague and Politics.' *Bull. Hist. Med*. Vol XXVIII 1954, p409.

33 cit. Hecker. p38.

34 p21 above.

101 35 cit. Nohl. p252.

36 Guillaume de Machaut. *Jugement du Roy de Navarre*.

37 S. Guerchberg. 'La controverse sur les prétendus semeurs de la Peste Noire.' *Revue des Études Juives*. N.S. Tome VIII 1948, pp3-40.

102 38 E. Wickersheimer. 'La Peste Noire à Strasbourg.' *Proc. 3rd Int. Cong. Hist. Med*. Antwerp 1923, p54.

39 Text of confessions quoted by Hecker. op. cit. pp70-74.

103 40 Matthew of Neueburg. op. cit. p262.

41 Heinricus de Diessenhoven. *Fontes Rerum Germanicarum*. Vol IV p68.

42 Michael Kleinlawel. *Strassburgische Chronik*. cit. Nohl. p242.

43 Heinrici Rebdorfensis. 'Annales Imperatorum.' *Fontes Rerum Germanicarum*. Vol IV p534.

106 44 'Aegidii Li Muisis.' De Smet. op. cit. Vol II p342-43.

106 45 See, in particular, R. Hoeniger, *Der Schwarze Tod in Deutsch-land*. Berlin 1882, pp9-11.

46 July 4 and Sept. 26, 1348. Raynaldus. *Annales eccles.* ed. Mansi. Vol VI 1750, p476.

47 Hecker. op. cit. p42.

107 48 Haeser. op. cit. Vol III p181.

49 J. Parkes. *The Jews in the Mediaeval Community*. London 1938, p118.

50 A. Lopez de Meneses. 'Una consecuencia de le Peste Negra en Cataluña: El pogrom de 1348.' *Sefarad* XIX 1959, p92.

51 'Documentos acerca de la Peste Negra en los dominios de la Corona de Aragon.' *Consejo superior de Investigaciones Cientificas.* Vol VI 1956, p298.

108 52 L. Bertrand. 'Contribution a l'Étude de la Peste dans les Flandres.' *Proc. 2nd Int. Cong. Hist. Med.* Evreux, 1922, p43.

CHAPTER SIX

111 1 C. 5. Bartsocas. *Journal of the History of Medicine*. Vol XXI No. 4 1966, p395.

2 op. cit. p13.

3 Farlati. *Illyricum Sacrum*. iii p324.

4 *Historia*. iii p406.

112 5 Gasquet. op. cit. p78.

6 Nohl. op. cit. p37.

113 7 op. cit. p28.

8 C. Verlinden's monograph, 'La Grande Peste de 1348 en Espagne' in the *Revue belge de Philologie et d'Histoire*, t. XVII 1938, p103, is the best general study yet published.

9 'Documentos acerca de la Peste Negra en los dominios de la Corona de Aragon.' op. cit. p291 and 'Una Consecuencia de la Peste Negra en Cataluña: El Pogrom de 1348.' op. cit. p92.

10 'Documentos acerca . . .' op. cit. April 20, 1348.

114 11 Philippe. *Histoire de la Peste Noire*. p54.

12 Sudhoff. *Archiv*. XIX pp46-48.

13 Walsingham, R.S. 28, I p273, cf. Capgrave, R.S. 1 p213. (This may not relate to Spain in particular though it could as well apply there.)

115 14 'Chronicon majus Aegidii Li Muisis.' De Smet. Vol II p280.

CHAPTER SEVEN

118   1   E. M. Carus Wilson. *Mediaeval Merchant Venturers.* London 1945, p240 et seq.

2   G. A. Holmes. *The Estates of the Higher Nobility in 14th Century England.* Cambridge 1957, p5.

3   E. B. Fryde. 'The Last Trials of Sir William de Pole.' *Econ. Hist. Rev.* Ser. Vol XV 1962, p17.

119   4   E. A. Kosminsky. *Studies in the Agrarian History of England.* Oxford 1956, p322-23.

5   J. C. Russell. *British Mediaeval Population.* Albuquerque 1948, p287.

6   This figure is far from uncontested. Bennett suggests it may have been as low as 5,000 but most authorities agree that it lost population heavily between 1348 and 1377 and the poll tax figure for the latter date (always an underestimate) was nearly 6,000.

7   'A 14th Century Chronicle from the Grey Friars at Lynn.' Ed. A. Grandsen. *Eng. Hist. Rev.* Vol LXXII 1957, p274.

8   *Chronica Monasterii de Melsa.* R.S. 43 III p68. See also Higden's *Polychronicon.* R.S. 41 VIII 355.

9   Knighton. op. cit. p61.

120   10   Capgrave. ed. F. C. Hingeston. R.S. 1 p213.

11   *Eulogium (Historiarum sive Temporis).* R.S. 9 III p213.

12   *Canon of Bridlington's Chronicle* (R.S. 76 II p149). *Galfridi le Baker* op. cit. p99.

13   *Continuatio Chronicarum.* R.S. 93 p406.

121   14   'Vitae Archiepiscoporum.' *Anglia Sacra.* Vol 1 p42.

15   *Originalia Roll.* 24 Ed. III, m.2. cit. Gasquet. p81.

122   16   *Studies in Agrarian History.* op. cit. p321.

123   17   *Revue belge de Philologie et d'Histoire.* t. XXVII 1950, p600.

18   op. cit. pp86-89.

124   19   J. M. Fletcher. 'The Black Death in Dorset.' *Dorset Nat. Hist. Ant. Field Club.* Vol XLIII 1922, p1.

20   *Hist. MSS. Comm.* 6th Report. p475.

21   Wilkins. *Concilia* ii pp735-36.

126   22   Dr. J. Lunn's Ph.D. thesis of 1930. Most unfortunately no copy of this survives but many of its valuable statistics are quoted in Dr. Coulton's *Mediaeval Panorama* (pp495-99 and notes).

23   Gasquet. op. cit. p96.

127   24   A. Hamilton Thompson. 'Pestilences of the 14th Century in the Diocese of York.' *Archaeological Journal.* Vol 71 1914, pp98-100.

25   op. cit. p192.

128   26   op. cit. p230.

129   27   'Register of Bishop Ralph of Shrewsbury.' *Somerset Record Society*. Vol X 1896, p596.

130   28   M. Baehrel. 'Epidémie et Terreur: Histoire et sociologie.' *Annales historiques de la Révolution française*. Vol XXIII 1951, pp113-46 and 'La haine de classe en temps d'épidémie.' *Annales E.S.C.* Vol VII No. 2 1952, pp 351-60.

132   29   *Victoria County History* (henceforth referred to as V.C.H.), Hampshire. Vol II p33. See p147 below.

133   30   *The Sky Suspended*. London 1960, p168.

135   31   Knighton. op. cit. p61.

      32   C. E. Boucher. 'The Black Death in Bristol.' *Transactions of the Bristol and Gloucestershire Archaeological Society*. Vol IX 1938, p36.

      33   S. Seyer. *Memoirs of Bristol*. Bristol 1823, Vol II p143.

      34   A. Jenkins. *History of the City of Exeter*. Exeter, 1841, p62.

      35   G. Oliver. *History of the City of Exeter*. Exeter 1861, p74.

136   36   W. G. Hoskins. *Devon*. London 1954, p169.

      37   Dr. J. Lunn. Ph.D. Thesis.

      38   L. F. Salzmann. *English Industries of the Middle Ages*. London 1913, p74. A. R. Bridbury. *Economic Growth*. London 1962, p25.

CHAPTER EIGHT

137   1   'Lives of the Berkeleys.' ed. J. Smyth. *Bristol and Gloucestershire Archaeological Society*. Gloucester 1883, Vol I p322.

138   2   *Galfridi le Baker*. op. cit. p99.

      3   *V.C.H. Gloucestershire*. Vol II p19.

      4   'Lives of the Berkeleys.' op. cit. Vol I p307.

139   5   A. Hamilton Thompson. 'Register of John Gynewell, Bishop of Lincoln, for the Years 1347-50.' *Archaeological Journal*. Vol 68 1911, p323 and App. 3.

      6   'Eynsham Cartulary.' ed. H. E. Salter. *Oxford Historical Society*. 1907-8 Vol 2 p69.

      7   M. Beresford. *Lost Villages of England*. London 1954, p159.

140   8   'Eynsham Cartulary.' Vol 2 p69. cf. K. J. Allison and other members of the Deserted Mediaeval Village Research Group. *The Deserted Villages of Oxfordshire*. Leicester 1965.

      9   P. D. A. Harvey. *A Mediaeval Oxfordshire Village: Cuxham*. Oxford 1965, p64.

      10 A. Wood. *History and Antiquities of the University of Oxford*. Oxford 1792, Vol 1 p449.

11   E. Brown. *Fasciculus rerum expetendarum et fugiendarum*. London 1690, Vol 2 p473.

12   *Loci e libro veritatum*. ed. J. E. T. Rogers. Oxford 1881, p202.

141   13   *De Ecclesia*. ed. J. Loserth. London 1886, p374.

14   H. E. Salter. *Mediaeval Oxford*. Oxford 1936, p108. cf. Hastings Rashdall. *Universities of Europe in the Middle Ages* (ed. Powicke and Emden). Oxford 1936, Vol 3 p317.

141   15   *V.C.H. Berkshire*. Vol II p185-87.

142   16   Hamilton Thompson. op. cit. p322.

17   L. J. Ashford. *History of the Borough of High Wycombe*. London 1960, p49.

143   18   *V.C.H. Wiltshire*. Vol IV p39.

144   19   Gasquet. op. cit. p130.

20   *Reg. Edendon* ii fol. 17. 'Mandatum ad orandum pro Pestilentia.' cit. Gasquet. op. cit. p124.

145   21   *V.C.H. Hampshire*. Vol II pp32-33.

22   Dr. J. Lunn. cit. Coulton. p496.

23   N. S. and E. C. Gras. *The Economic and Social History of an English Village*. Harvard 1930, p153.

146   24   Ibid. p76.

25   Gasquet, op. cit. pp216-18.

26   *Originalia Roll*. 29, Ed. III m. 8. cit. Gasquet. p217.

27   *British Mediaeval Population*. op. cit. p285.

147   28   W. L. Woodland. *The Story of Winchester*. London 1952, p114. *V.C.H. Hampshire*. Vol 11 p32.

148   29   H. C. M. Lambert. *History of Banstead in Surrey*. Oxford 1931, p15.

30   E. Robo. 'The Black Death in the Hundred of Farnham.' *Eng. Hist. Rev.* Vol XLIV 1929, p560.

31   See p226 below.

CHAPTER NINE

151   1   J. C. Russell. *British Mediaeval Population*. op. cit. pp286-87.

153   2   I have made much use of E. L. Sabine's three essays in *Speculum*: 'Butchering in Mediaeval London,' Vol VIII 1933, p335; 'Latrines and Cess-pools of Mediaeval London,' Vol IX 1934, p303 and 'City cleaning in Mediaeval London,' Vol XII 1937, p19, in preparing this chapter.

154   3   B. Lambert. *History and Survey of London*. London, 1806 Vol 1 p241.

155   4   H. J. Riley. *Memorials of London and London Life*. London 1868, p295.

156   5   'Historical MSS belonging to the Dean and Chapter of Canterbury.' *H. Mss. Comm.* Second Report. p338.

       6   Robert of Avesbury. R.S. 93 p407.

       7   Greenwood. *Epidemics and Crowd Diseases.* op. cit. p291.

157   8   McKisack. *The Fourteenth Century.* Oxford 1949, p220.

       9   Dom. D. Knowles. *The Religious Orders in England.* Cambridge 1955, Vol II pp130-31.
W. Hope. *History of the London Charterhouse.* London 1925, p8.

    10   op. cit. p407.

158 11   *Survey of London.* Vol II p81.

    12   *Abstract of the Population Returns of 1831.* London 1832, p11.

159 13   C. H. Talbot and E. A. Hammond. *The Medical Practitioner in Mediaeval England.* London 1965, p312.

    14   Creighton, op. cit. Vol I p129.

    15   *Chronicon Johannis de Reading.* ed. J. Tait. Manchester 1914, p108.

    16   A. R. Stanley. *Memorials of Westminster Abbey.* London 1868, pp376-77.

160 17   J. C. Russell. op. cit. p285.

    18   Creighton. op. cit. Vol I p195.

    19   Knighton, op. cit. p120.

    20   John of Reading. op. cit. pp109-10.

CHAPTER TEN

161   1   *V.C.H. Sussex.* Vol II p77.

      2   Ibid. Vol II p54.

      3   Ibid. Vol II p182.

      4   Willelmi de Dene. 'Historia Rossensis.' Wharton. *Anglia Sacra.* Vol I pp375-76.

163   5   J. E. T. Rogers. *Six Centuries of Work and Wages.* London 1906, p221.

      6   C. E. Woodruff and W. Danks. *Memorials of Canterbury Cathedral.* London 1912, p148.

      7   See p78 above.

      8   J. C. Russell. op. cit. p216.

164   9   Stephani Birchington. op. cit. p42.

    10   J. E. T. Rogers. op. cit. p225.

    11   A. Hamilton Thompson. 'Registers of John Gynewell.' op. cit. p322.

165 12   C. R. Haines. *Dover Priory.* Cambridge 1930, p267n.

    13   A. E. Levett. *Studies in Manorial History.* Oxford 1938, p251 et seq.

166 14 E. Toms. *The Story of St. Albans.* St. Albans 1962, p50-51.

15 *Gesta Abbatum S. Albani.* R.S. 28 Vol ii p369. cf. L. F. R. Williams. *History of the Abbey of St. Albans.* London 1917, p166.

16 A. Hamilton Thompson. op. cit. p324.

17 *V.C.H. Bedford.* Vol III p318.

167 18 A. Jessop. 'The Black Death in East Anglia.' *The Coming of the Friars and other Historic Essays.* London 1894, p200-1.

19 C. G. Grimwood. *History of Sudbury.* Sudbury 1952, p86.

20 F. Seebohm. 'The Black Death and its place in English History.' *Fortnightly Review.* Vol II 1865, p155.

21 Gasquet. op. cit. p153.

168 22 G. A. Holmes. *The Estates of the Higher Nobility* . . . op. cit. p90 et seq.

23 Ibid. p115.

169 24 K. J. Allison. 'The Lost Villages of Norfolk.' *Norfolk Archaeology.* Vol XXXI 1955, p131.

25 *V.C.H. Suffolk.* Vol II p19.

170 26 F. Blomefield. *History of the County of Norfolk.* Vol III London 1806.

27 F. Seebohm. 'The Black Death and its place in English History.' *Fortnightly Review.* Vol II 1865, p157-58.

28 J. C. Russell. *British Mediaeval Population.* op. cit. p293.

29 *V.C.H. Suffolk.* Vol II p19.

30 op. cit. p206.

171 31 F. R. Chapman. *Sacrist Rolls of Ely.* Cambridge 1907, p107.

32 Gasquet. op. cit. p154.

33 *V.C.H. Cambridgeshire.* Vol II p158.

172 34 *Hist. MSS. Comm.* 6th Report. App. p299.

35 Jessop, op. cit. p220.

CHAPTER ELEVEN

174 1 *V.C.H. Huntingdon.* Vol II p123.

175 2 A. Hamilton Thompson. op. cit. pp323-24.

3 A. Rogers. *The Making of Stamford.* Leicester 1965, p49.

4 Knighton. op. cit. pp61-62.

176 5 M. W. Beresford. *Lost Villages of England.* London 1954, p161.

6 Knighton. op. cit. p61.

7 C. J. Billson. *Mediaeval Leicester.* Leicester 1920, p143.

177 8 Ibid. p144-45.

178 9 Cox. *Notes on the Churches of Derbyshire.* P. VIII.

10 A. Hamilton Thompson. 'The Pestilences of the 14th century in the Diocese of York.' *Archaeological Journal.* Vol 71 1914, pp111-12.

178   11   *Fortnightly Review*. Vol II 1865, p151.

      12   *Black Death*. p173.

      13   A. Hamilton Thompson. Lincoln. op. cit. p326.

179   14   I. W. F. Hill. *Mediaeval Lincoln*. Cambridge 1948, p252.

      15   Ibid. p251.

      16   *Chronicle of Louth Park*. Lincolnshire Record Society. 1891, pp38-39.

      17   M. W. Beresford. *Lost Villages of England*. op. cit. p203.

      18   H. E. Hallam. 'Population Density in Mediaeval Fenland.' *Econ. Hist. Rev.* 2nd Series, Vol XIV 1961, No 1 p78

180   19   *Historical Papers from Northern Registers*. R.S. 61 p395-97.

182   20   Dr. Lunn. p126, n22 above.

      21   A. Hamilton Thompson. *York*. op. cit. p107-8.

      22   Ibid. p110.

      23   J. C. Russell. op. cit. p142.

      24   J. N. Bartlett. 'The expansion and decline of York in the later Middle Ages.' *Econ. Hist. Rev.* 2nd Ser. Vol XII 1959, p17.

      25   C. B. Knight. *History of the City of York*. York 1944, p222.

      26   M. E. Jeanselme. 'Inondations, Famines et Tremblements de Terre sont les avant-coureurs de la Peste.' *Proc. 3rd Int. Cong. Hist. Med.* (1922).

183   27   *Chronica Monasterii de Melsa*. R.S. 43 III p69.

      28   Ibid. p37.

      29   T. Blashill. *Sutton in Holderness*. op. cit. p98.

      30   T. Burton. *The History and Antiquities of the Parish of Hemingborough*. York 1888, p271.

      31   Gasquet. op. cit. p181.

184   32   H. Fishwick. *History of Lancashire*. London 1894, p74.

185   33   R.S. France. 'A History of Plague in Lancashire.' *Trans. Hist. Soc. of Lancs and Cheshire*. Vol 90 1938, p24.

      34   Gasquet. op. cit. pp183-84.

186   35   *V.C.H. Durham*. Vol II p210.

      36   Surtees. *History of Durham*. Vol I p.lii.

      37   *V.C.H. Durham*. Vol II p212.

CHAPTER TWELVE

187   1   Gasquet. op. cit. p141.

      2   *V.C.H. Worcestershire*. Vol II p32.

      3   T. R. Nash. *History of Worcestershire*. London 1781, Vol I p226.

      4   V. Green. *History of Worcester*. London 1796, p144.

188   5   p129 above.

      6   H. L. V. Fletcher. *Herefordshire*. London 1948, p22.

189 7 Owen and Blakeway. *History of Shrewsbury*. London 1825, Vol 1 p165.

8 Gasquet. op. cit. p170.

190 9 W. Rees. 'The Black Death in England and Wales, as exhibited in Manorial Documents.' *Proc. Roy. Soc. Med.* Vol. XVI Pt. 2 p34.

191 10 *Galfridi le Baker*. op. cit. p100.

11 The subsequent paragraphs draw heavily on W. Rees's monograph 'The Black Death in Wales.' *Trans. Roy. Hist. Soc.* Fourth Series. Vol III 1920.

192 12 *Court Rolls*. Portfolio 218. No 4 cit. Rees.

195 13 Friar John Clyn. *Annals of Ireland*. ed. R. Butler. Irish Arch. Soc. Dublin 1849, p37.

14 I am fortunate in having been able to consult in proof Chapter VIII of Dr. Otway Ruthven's *History of Mediaeval Ireland* (London 1968). A. Gwynn's monograph 'The Black Death in Ireland' (*Studies: An Irish Quarterly Review.* Vol XXIV 1935, pp25-42) is also of value.

15 A. Gwynn. op. cit. p28.

196 16 *Annals of Connacht*. ed. A. M. Freeman. Dublin Institute for Advanced Studies. 1944. cit. Ruthven.

197 17 *Galfridi le Baker*. op. cit. p100.

198 18 op. cit. pp62-63.

19 Col. MacArthur. 'Old Time Plague in Britain.' *Trans. Roy. Soc. Trop. Med. Hyg.* Vol XIX p360.

20 *Chronicle of the Scottish Nation*. ed. W. F. Skene. Edinburgh 1872.

199 21 *Cronykil of Andrew of Wyntoun*. ed. D. Laing, Edinburgh 1872, Vol II p482.

22 *David Macpherson's preface to 1795 edition of the Chronicle.* London. pXVII.

23 ed. F. J. Skene. Edinburgh 1880, p225.

CHAPTER THIRTEEN

In writing this chapter I have found of particular value:

R. H. Hilton.	*A Mediaeval Society*. London 1966.
H. S. Bennett.	*Life on the English Manor*. Cambridge 1956.
A. Jessop.	*The Coming of the Friars and other Historic Essays*. London 1894.
J. J. Jusserand.	*English Wayfaring life in the Middle Ages.* London 1891.
G. G. Coulton.	*Mediaeval Panorama*. Cambridge 1938.
G. C. Homans.	*English Villagers of the Thirteenth Century.* Cambridge (Mass) 1942.

and, analysing the effect of the Black Death on a village or group of villages:

P. D. A. Harvey.	*A Mediaeval Oxfordshire Village: Cuxham.* Oxford 1965.
A. E. Levett.	*The Black Death on the Estates of the See of Winchester.* Oxford 1916.
E. Robo.	*The Black Death in the Hundred of Farnham.*

Some of these relate to a period somewhat before the Black Death. Others have had to be used with discretion because they deal with areas of England other than that where Blakwater is situated. But the overall picture has not been falsified.

CHAPTER FOURTEEN

224 1   Though see J. C. Russell. *British Mediaeval Population.* op. cit. p54.

225 2   J. Z. Titow. 'Some evidence of the 13th century population increase.' *Econ. Hist. Rev.* 2nd Ser. Vol XIV No 2 1961, p220.

3   M. Postan. 'Some economic evidence of declining population in the later Middle Ages.' *Econ. Hist. Rev.* 2nd Ser. Vol II 1950, p221.

4   F. Seebohm. 'The Black Death and its Place in English History.' *Fortnightly Review.* Vol II 1865, pp149-60 and 268-79.

5   J. E. T. Rogers. 'England before and after the Black Death.' *Fortnightly Review.* Vol III 1865, pp191-96.

6   F. Seebohm. 'The Population of England before the Black Death.' *Fortnightly Review.* Vol IV 1866, pp87-89.

226 7   J. C. Russell. op. cit. p246.

8   Ibid. pp22-33.

9   G. C. Homans. *English Villagers of the Thirteenth Century.* op. cit. pp209-12.

10   J. Z. Titow. op. cit. p222.

11   J. Krause. 'The Mediaeval Household: Large or Small.' *Econ. Hist. Rev.* 2nd Ser. Vol IX 1957, p432.

227 12   G. R. Elton. *The Practice of History.* Sydney 1967, p34.

13   J. C. Russell. 'Recent Advances in Mediaeval Demography.' *Speculum.* Vol XL No 1 1965, p84.

14   *Black Death.* op. cit. p225.

15   pp127-28 above.

228 16   'Registers of the Bishop of Lincoln' and 'Pestilences of the 14th century in the Diocese of York.' op. cit.

17   PhD. Thesis. op. cit. p126, n22, above.

18   p128 above.

19 op. cit. p221.

229 20 E. Robo. 'The Black Death in the Hundred of Farnham.' *Eng. Hist. Rev.* Vol XLIV 1929, p560.

21 P. D. A. Harvey. *A Mediaeval Oxfordshire village: Cuxham.* op. cit. p135.

22 F. M. Page. *The Estates of Crowland Abbey.* Cambridge 1934, p125.

23 A. E. Levett. 'The Black Death on the Estates of the See of Winchester.' *Oxford Studies in Social and Legal History.* Vol V Oxford 1916, p80-81.

24 op. cit. p216.

230 25 Ibid. p367.

231 26 Y. Renouard. 'Conséquences et intérêt démographique de la Peste Noire de 1348.' *Population.* t. III 1948, p459.

27 A. Doren. *Storia Economica dell' Italia nel Medio Evo.* Padua 1937, p579.

CHAPTER FIFTEEN

232 1 E. R. Hume. *History of England.* Vol II London 1796, p448.

2 R. Henry. *The History of Great Britain.* Vol VII London 1788, p246.

3 J. R. Green. *History of the English People.* Vol 1 London 1877, pp429-30.

4 E. Friedell. *Kulturgesicht der Neuzeit.* Vol 1 Munich 1927, p62.

5 G. M. Trevelyan. *English Social History.* London 1942, P.XI.

233 6 J. Thorold Rogers. *A History of Agriculture and Prices in England.* Vol 1 Oxford 1866.

234 7 Greenwood. *Epidemics and Crowd Disease.* op. cit. p291. cf. J. M. W. Bean. 'Plague, Population and Economic Decline in England in the later Middle Ages.' *Econ. Hist. Rev.* 2nd Ser. Vol XV 1963, pp427-28.

8 R. S. Roberts. 'The Place of Plague in English History.' *Proc. Roy. Soc. Med. (Hist. Med.)* Vol 59 1966, p101.

235 9 Thorold Rogers, for instance, argued that England could not have supported a population of more than 2½ million. *Fortnightly Review.* Vol II 1865, pp191-96.

10 P. Vinogradoff. Review of *'The End of Villainage in England'* by T. W. Page. *Eng. Hist. Rev.* Vol. XV 1900, p776.

11 A. E. Levett. *The Black Death on the Estates of the See of Winchester.* Oxford 1916, p63.

236 12 *Black Death.* op. cit. p46.

13 J. T. Rogers. op. cit. p265.

236 14 P. D. A. Harvey. *A Mediaeval Oxfordshire Village*; *Cuxham*. Oxford 1965, App. IV.

     15 *V.C.H. Middlesex*. Vol II p80.

     16 J. T. Rogers. op. cit. Chapter XV.

     17 Levett. op. cit. p100.

237 18 W. Beveridge 'Wages in the Winchester Manors.' *Econ. Hist. Rev.* Vol VII 1936/37, p26.

     19 W. Beveridge. 'Westminster wages in the Manorial Era.' *Econ. Hist. Rev.* 2nd Ser. Vol VIII 1955, p18.

     20 Knighton. op. cit. p62.

     21 J. T. Rogers. op. cit. passim.

     22 E. Robo. op. cit. p149 above.

     23 *Eynsham Cartulary*. op. cit. p140 above.

238 24 F. G. Davenport. *The Economic Development of a Norfolk Manor*. Cambridge 1906, pp70-72.

     25 E. Lipson. *Economic History of England*. Vol 1 London 1945, p92.

     26 B. H. Putnam. *Enforcement of the Statute of Labourers*. New York 1908, p91.

     27 Ibid. p223.

239 28 G. A. Holmes. *The Estates of the Higher Nobility* . . . op. cit. pp90-92.

     29 M. McKisack. *The Fourteenth Century*. op. cit. p324.

240 30 T. W. Page. 'The End of Villainage in England.' *Publications of the American Economic Association*. 3rd Ser. Vol 1 1900, p39.

241 31 F. Pollock and F. W. Maitland. 'History of English Law.' *Eng. Hist. Rev.* Vol I p166.

     32 K. G. Feiling. 'An Essex Manor in the 14th Century.' *Eng. Hist. Rev.* Vol XXVI 1911, p333.

     33 H. L. Gray. 'The Commutation of Villain Services in England before the Black Death.' *Eng. Hist. Rev.* Vol XXIX 1914, p625.

242 34 A. E. Levett. op. cit. pp159-60.

     35 A. Ballard. *The Manors of Witney, Brightwell and Downton*. Oxford 1916, pp181-204.

     36 J. A. Raftis. *Estates of Ramsey Abbey*. Toronto 1957, p251.

243 37 P. D. A. Harvey. op. cit. p85.

     38 M. Postan. 'The Chronology of Labour Services.' *Trans. Roy. Hist. Soc.* 4th Ser. Vol XX 1937, pp185-86.

244 39 *V.C.H. Middlesex*. Vol. II p80.

     40 J. T. Rogers. op. cit. Chapter XV.

245 41 pp235-36 above.

246 42 *Eulogium (Historiarum sive Temporis)*. ed. F. S. Haydon. R.S. 9 III pp213-14.

     43 P. Vinogradoff. *Eng. Hist. Rev.* Vol. XV 1900, p779.

247 44 Knighton. op. cit. p64.
248 45 B. H. Putnam. op. cit. passim
46 L. F. Salzmann. *English Industries of the Middle Ages*. London 1913. .
249 47 A. E. Levett. op. cit. p134.
48 A. Reville. *Le Soulèvement des Travailleurs d'Angleterre en 1381*. Paris 1898. Introduction by C. Petit-Dutaillis, pXXXVII.
49 R. H. Hilton. 'Peasant Movements in England before 1381.' *Econ. Hist. Rev.* 2nd Ser. Vol II 1949, p117.
250 50 A. E. Levett. op. cit. p159.
51 P. Vinogradoff. *Eng. Hist. Rev.* Vol XV 1900, p779.
52 e.g. F. Lutge. 'Das 14/15 Jahrhundert in der Sozial und Wirtschaftsgeschichte.' *Jahrbücher f. Nationalökonomie und Statistik*. Vol. 162 1950, pp161-213; and E. Kelter. 'Das deutsche Wirtschaftsleben des 14 und 15 Jahrhunderts in Schatten der Pestepidemien.' *Ibid*. Vol 165 1953, pp161-208.

CHAPTER SIXTEEN

252 1 Campbell. op. cit. p162 (Grenoble, Vercelli, Reggio and Naples).
2 A. F. Leach. *The Schools of Medieval England*. London 1915, p197.
254 3 W. A Pantin. *The English Church and the Continent: The Later Middle Ages*. London 1959, p3.
255 4 Hastings Rashdall. *The Universities of Europe in the Middle Ages* (ed. Powicke and Emden). Oxford 1936, Vol III p317.
5 *Hist. MSS. Comm.* Vth Report App. (1874) p450.
6 *V.C.H. Oxfordshire*. Vol III p154.
7 J. E. T. Rogers. *Six Centuries of Work and Wages*. op. cit. p224.
256 8 E. Power. *The Wool Trade in English Mediaeval History*. Oxford 1941, p35.
257 9 E. Prior. *Cathedral Builders*. London 1905, p130.
10 Prior and Gardner. *Medieval Figure Sculpture in England*. London 1912, p390.
11 R. Crawfurd. *Plague and Pestilence in Literature and Art*. op. cit. pp130-31.
258 12 J. Harvey. *Gothic England*. London 1947, p40.

CHAPTER SEVENTEEN

261   1   Wilkins. *Concilia* ii pp735-36.

    2   Willelmi de Dene. op. cit. Vol I p375.

    3   Stephen Birchington. op. cit. p42.

262   4   *Harl M.S.* 6965. fol. 145.

    5   Gasquet. op. cit. p239.

    6   *Historical Papers from Northern Registers.* R.S. 61 p401.

    7   Knighton. op. cit. p63.

263   8   *Black Death.* op. cit. pp39-41.

    9   p127 above.

   10   Knighton. op. cit. p63.

264   11   Gasquet. op. cit. pp247-48.

   12   Ibid. p248.

265   13   D. Knowles. *The Religious Orders in England.* Vol II Cambridge 1955, p256-57.

   14   R. H. Snape. *English Monastic Finances.* Cambridge 1926, pp21-22.

   15   Hamilton Thompson. 'Gynewell . . .' op. cit. p328-29.

   16   P. Mode. *Influence of the Black Death on English Monasteries.* Chicago 1916.

266   17   C. F. Mullett. *The Bubonic Plague and England.* Lexington 1956, p34.

   18   *Annales Minorum.* Vol VIII p22.

267   19   Lea. op. cit. Vol I p290.

268   20   E. Carpentier. *Une ville devant la peste.* op. cit. p193.

   21   e.g. *Memorials of Canterbury Cathedral.* op. cit. p148.

   22   Matteo Villani. *Cronica* Florence 1846. Book 1 Chap. VII p15.

   23   M. Meiss. *Painting in Florence and Siena after the Black Death.* Princeton 1951, p79.

269   24   Ibid. p73.

270   25   *Black Death.* p74.

   26   op. cit. Book 1 Chap. IV p13.

271   27   R. Hoeniger. *Der Schwarze Tod in Deutschland.* op. cit. p133.

272   28   E. Carpentier. op. cit. pp195-96.

   29   J. W. Thompson. 'The Aftermath of the Black Death and the Aftermath of the Great War.' *American Journal of Sociology.* Vol XXVI 1920/21, p565.

   30   'La Peste Noire.' *Revue de Paris.* March, 1950, p117.

273   31   G. Prat. 'Alibi et la Peste Noire.' *Annales du Midi.* t. LXIV 1952, p15.

   32   J. C. Russell. 'Effects of Pestilence and Plague, 1315-85.' *Comparative Studies in Society and History.* Vol VIII No 4 1966, pp464-70.

274 33 S. Thrupp. 'Plague effects in Mediaeval Europe.' *Ibid.* pp482-83.

34 H. Baron. 'Franciscan Poverty and Civic Wealth.' *Speculum.* Vol XIII 1938, p12.

275 35 H. H. Mollaret and Jacqueline Brossolet. *La Peste, Source Méconnue d'Inspiration Artistique.* Paris. (Institut Pasteur), 1965, p60.

36 For a perceptive appreciation of this picture see P. Perdrizet. *La Peinture Religieuse en Italie jusqu'à la fin du XIVe siècle.* Nancy 1905, p47.

276 37 P. Perdrizet. *La Vierge de Miséricorde.* Paris 1908, p151.

38 Émile Mâle. *L'Art Religieux de la Fin du Moyen Age.* Paris 1908, p75.

277 39 See note 29 above.

279 40 J. J. Jusserand. *English Wayfaring Life in the Middle Ages.* London 1891, pp382-83.

Bibliography

There are remarkably few full-length studies dealing with the Black Death as a whole or even in a country or group of countries. The most important of these is still that by Cardinal Gasquet though many of his facts have now been disproved and his conclusions shown to be invalid. Sticker's study gives the widest coverage for Europe as a whole and Hoeniger's for Germany. The others are of slight importance.

COULTON, G. G.   *The Black Death*. London 1929.

GASQUET, F. A.   *The Great Pestilence*. London 1893. Reprinted substantially unrevised as *The Black Death*. London 1908.

HECKER, J. F. C.   *The Epidemics of the Middle Ages*. trad. Babington, London 1859.

HOENIGER, R.   *Der Schwarze Tod in Deutschland*. Berlin 1882.

LECHNER, K.   *Das Grosse Sterben in Deutschland*. Innsbrück 1884.

NOHL, J.   *Der Schwarze Tod*. Potsdam 1924.

PHILIPPE, A.   *Histoire de la Peste Noire*. Paris 1853.

STICKER, G.   *Die Pest*. Vol 1 ('Die Geschichte der Pest'). Giessen 1908.

More useful material on a national or international scale is often to be found in books not dealing exclusively with the Black Death (Coulton's *Mediaeval Panorama*, for instance, contains more of value than his monograph mentioned above) or in more recent essays and articles. In this and subsequent sections I have marked with an asterisk sources of particularly valuable information.

*CARPENTIER, E.   'Autour de la Peste Noire.' *Annales E.S.C.* 1962, t. XVII p1062.

*COULTON, G. G.   *Mediaeval Panorama*. Cambridge 1938, Chapter 38.

DOREN, A.   *Storia Economica dell' Italia nel Medio Evo*. Padua 1937.

DUBY, G.   *L'Économie rural et la vie des campagnes dans L'Occident médiéavl*. Paris 1962.

GWYNN, A.   'The Black Death in Ireland.' *Studies*. 1935 Vol XXIV p25.

MAYCOCK, A. L.   'A Note on the Black Death.' *Nineteenth Century*. 1925 Vol XCVII p456.

*REES, W.   'The Black Death in Wales.' *Trans. Roy. Hist. Soc.* Fourth Series 1920 Vol III p115.

REES, W.   'The Black Death in England and Wales as exhibited in Manorial Documents.' *Proc. Roy. Soc. Med.* Vol 16 Pt 2 p27.

*RENOUARD, Y.   'La Peste Noire.' *Revue de Paris*. March 1950, p107.

ROGERS, J. E.   'England before and after the Black Death.'
THOROLD   *Fortnightly Review*. 1865 Vol III p191.

RUTHVEN, O.   *History of Medieval Ireland*. London 1968.

SEEBOHM, F.   'The Black Death, and its Place in English History.' *Fortnightly Review*. 1865 Vol II pp149 and 268.

SEEBOHM, F.   'The Population of England before the Black Death.' *Fortnightly Review*. 1866 Vol IV p87.

VERLINDEN, C.   'La Grande Peste de 1348 en Espagne.' *Revue belge de Philologie et d'Histoire*. 1938 t XVII p103.

Among works on epidemiology, medical history or bubonic plague, those of particular relevance to the Black Death are:

ANGLADA, A.   *Étude sur les Maladies Éteintes*. Paris 1869.

CREIGHTON, C.   *A History of Epidemics in Britain*. Cambridge 1891.

*GREENWOOD,   *Epidemics and Crowd Diseases*. London 1935.
MAJOR

*HIRST, L. F.   *The Conquest of Plague*. Oxford 1953.

JOHN, F. M.   *The Black Death*. London 1920.

LISTON, W. G.   'The Plague.' *Brit. Med. Journ.* 1924 Vol 1 pp900 950 and 997.

MACARTHUR, W.   'Old Time Plague in Britain.' *Trans. Roy. Soc. Trop. Med. Hyg.* Vol XIX p355.

MULLETT, C. F.   *The Bubonic Plague and England*. Lexington 1956.

PAPON, J. P.   *De la Peste ou Époques Mémorables de ce Fléau*. Paris 1800.

*POLLITZER, R.   *Plague*. W.H.O. Geneva 1954.

REBOUIS, H. E.   *Étude historique et critique sur la peste*. Paris 1888.

SINGER, C.   'A Review of the Medical Literature of the Dark Ages.' *Proc. Roy. Soc. Med. (Hist. Med.)* Vol 10 Pt 2 p107.

ZINSSER, H.   *Rats, Lice and History*. London 1935.

Innumerable studies exist dealing in whole or in part with the Black Death or its effects in specific towns or areas. Some of these, for instance Dr. Carpentier's study of Orvieto, are of the greatest importance; others contain little except an odd anecdote or two and some inaccurate statistics. All those cited below have contributed something of value to this book. The Victoria County Histories, though varying greatly in quality from county to county, are in general a source of much valuable material for England.

ALLISON, K. J.   'The Lost Villages of Norfolk.' *Norf. Arch.* Vol XXXI 1955, p118.

BALLARD, A.   'The Manors of Witney, Brightwell and Downton.' *Oxford Studies in Social and Legal History.* Vol V. Oxford 1916.

BARTLETT, J. N.   'The Expansion and Decline of York in the Later Middle Ages.' *Econ. Hist. Rev.* 2nd Ser Vol XII 1959, p17.

BERTRAND, L.   'Contribution à L'Étude de la Peste dans les Flandres.' *Proc. 2nd. Int. Cong. Hist. Med.* Evreux 1922, p43.

BEVERIDGE, W.   'Wages in the Winchester Manors.' *Econ. Hist. Rev.* 1936-37 Vol VII p22.

BEVERIDGE, W.   'Westminster Wages in the Manorial Era.' *Econ. Hist. Rev.* 2nd Series, Vol VIII 1955 No 1 p18.

BILLSON, C. J.   *Mediaeval Leicester.* Leicester 1920.

BOUCHER, C. E.   'The Black Death in Bristol.' *Trans. Bristol and Glos. Arch. Soc.* Vol LX 1938.

BOWSKY, W. M.   'The Impact of the Black Death upon Sienese Government and Society.' *Speculum.* Vol XXXIX 1964 No 1 p1.

BRUNETTI.   'Venezia durante la Peste.' *Ateneo Veneto.* 32 1909.

BUESS, H.   'Die Pest in Basel im 14 und 15 Jahrhundest.' *Basel Jahrbuch* 1956.

*CARPENTIER, E.   *Une Ville devant la Peste. Orvieto et la Peste Noire de 1348.* Paris 1962.

*CHIAPPELLI, A.   'Gli Ordinamenti Sanitari del Comune de Pistoia contra la Peste de 1348.' *Arch. Stor. Ital.* Ser IV Vol XX p3.

DAVENPORT, F.   *The Economic Development of a Norfolk Manor, 1086-1565.* Cambridge 1906.

DUBLED, H.   'Aspects économiques de la vie de Strasbourg aux 13ᵉ et 14ᵉ siècles, "*Archives de L'Église d'Alsace.*" N.S. t. VI, 1955, No. 1, p. 18.

DUBLED, H.   'Conséquences économiques et sociales des "mortalitiés" du XIVᵉ siècle essentiellement ne Alsace.' *Revue d'Hist. Écon. et Soc.* Vol XXXVII 1959 No 3 p273,

EMERY, R.   'The Black Death of 1348 in Perpignan.' *Speculum.* Vol XLII 1967 No 4 p611.

FEILING, K. G.   'An Essex Manor in the 14th Century.' *Eng. Hist. Rev.* Vol XXVI 1911, p333.

FISHER, J. L. 'The Black Death in Essex.' *Essex Review.* Vol LII 1943.

FLETCHER, J. M. 'The Black Death in Dorset.' *Dorset Nat. Hist. Ant. Field Club.* Vol XLIII 1922, p1.

FRANCE, R. S. 'A History of Plague in Lancashire.' *Trans. Hist. Soc. Lancs and Cheshire.* Liverpool 1938 Vol 90 p1.

*HAMILTON THOMPSON, A. 'The Pestilences of the 14th Century in the Diocese of York.' *Arch. Journ.* Vol 71 1914, p97.

*HAMILTON THOMPSON, A. 'Registers of John Gynewell, Bishop of Lincoln, for the years 1347-50.' *Arch. Journ.* Vol 68 1911, p302.

HARVEY, P. D. A. *A Mediaeval Oxfordshire Village: Cuxham.* Oxford 1965.

*HERLIHY, D. 'Population, Plague and Social Change in Rural Pistoia.' *Econ. Hist. Rev.* 2nd Series 1965 Vol XVIII No 1 p225.

HEWITT, H. J. *Mediaeval Cheshire.* Manchester 1929.

HILL, I. W. F. *Mediaeval Lincoln.* Cambridge 1948.

HILTON, R. H. *The Economic Development of Some Leicestershire Estates in the 14th and 15th Centuries.* Oxford 1947.

HOSKINS, W. G. *Devon.* London 1954.

JESSOP, A. 'The Black Death in East Anglia' *The Coming of the Friars and other Historic Essays.* London 1894.

*LEVETT, A. E. 'The Black Death on the Estates of the See of Winchester.' *Oxford Studies in Social and Legal History.* Vol V Oxford 1916.

*LOPEZ DE MENESES, A. 'Documentos acerca de la peste negra en los dominios de la Corona de Aragon.' *Consejo Superior de Investigaciones Cientificas. Escuela de Estudios Mediaevales.* Vol VI 1956, p291.

LOPEZ DE MENESES, A. 'Una Consecuencia de la Peste Negra en Cataluña: El Pogrom de 1348.' *Sefarad.* Vol 19 1959, p92.

MOLLAT, M. 'La Mortalité à Paris.' *Moyen Age.* Vol 69 1963, p505.

NATHAN, M. *The Annals of West Coker.* Cambridge 1957.

PAGE, F. M. *The Estates of Crowland Abbey.* Cambridge 1934.

PORQUET, L. *La Peste en Normandie.* Vire 1898.

PRAT, G. 'Albi et la Peste Noire.' *Annales du Midi.* t. LXIV 1952, p15.

RAFTIS, J. A. *Estates of Ramsey Abbey.* Toronto 1957.

REINCKE, H. 'Bevölkerungsverluste der Hansestädte durch den Schwarzen Tod.' *Hansische Geschichtsblätter.* Vol 72 1954, p88.

RILEY, H. T. *Memorials of London and London Life.* London 1868.

ROBO, E. 'The Black Death in the Hundred of Farnham.' *Eng. Hist. Rev.* Vol XLIV 1929, p560.

ROUCAUD, J.   *La Peste à Toulouse.* Toulouse 1918.
SAUNDERS, H. W.   *An Introduction to the Obedientiary and Manor Rolls
                     of Norwich Cathedral Priory.* Norwich 1930.
SMYTH, J.   'The Lives of the Berkeleys.' *Bristol and Glos. Arch.
                     Soc.* Gloucester 1883.
WICKERSHEIMER,   'La Peste Noire à Strasbourg et le Régime des
E.   cinq médecins strasbourgeois.' *Proc. 3rd. Int.
                     Cong. Hist. Med.* Antwerp 1923, p54.
WILLIAMSON, R.   'The Plague in Cambridge.' *Med. Hist.* 1957, I (1)
                     p51.

It would be absurd to attempt to list all the general works on the period
which have contributed something to this book; equally it would be
churlish not to mention at least:

BENNETT, H. S.   *Life on the English Manor.* Cambridge 1956.
BRIDBURY, A. R.   *Economic Growth: England in the later Middle Ages.*
                     London 1962.
CAMBRIDGE ECONOMIC HISTORY. Vols 1 to 4. In particular Pro-
                     fessor Postan's and Professor Glenicot's contribu-
                     tions to the second edition (1966) of Vol I.
CAMBRIDGE MEDIAEVAL HISTORY. Vol III. Cambridge 1932.
HUIZINGA, J.   *The Waning of the Middle Ages.* London 1924.
JUSSERAND, J. J.   *English Wayfaring Life in the Middle Ages.* London
                     1891.
KNOWLES,   *The Religious Orders in England.* Vol II. Cambridge
DOM. D.   1955.
LEA, H. C.   *A History of the Inquisition of the Middle Ages.*
                     New York 1887-88.
LEVETT, A. E.   *Studies in Manorial History.* Oxford 1938.
LIPSON, E.   *The Economic History of England.* London 1945.
MᶜKISACK, M.   *The Fourteenth Century.* Oxford 1959.
PIRENNE, H.   *Economic and Social History of Mediaeval Europe.*
                     Trans. I. E. Clegg. London 1936.
ROGERS, J. E.   *A History of Agriculture and Prices in England.*
THOROLD   Vol 1. Oxford 1866.
                     Six Centuries of Work and Wages. London 1906.
SIMONDE DE   *Histoire des Républiques Italiennes du Moyen Âge.*
SISMONDI, J. C.   Vol VI, Paris 1826.

There are a large number of additional studies which illuminate one
aspect or another of the Black Death. It would in some ways have
been desirable to sub-divide them further into groups such as 'economic
effects' or 'anti-semitism' but this could only be done at the expense of
a comprehensive and easily consulted list of authors and I have therefore decided to lump all the remaining texts, except those of the

contemporary chroniclers and tractators, in a single category. This list does not contain books, cited in the notes, which deal with the Black Death only indirectly yet contribute something to the general picture.

ALLYN, H. M.   'The Black Death. Its Social and Economic Results.' *Annals of Medical History.* Vol VII 1925, p226.

BAEHREL, M.   'Épidémie et Terreur: Histoire et Sociologie.' *Annales Hist. de la Rev. Fran.* t. XXIII 1951, p113.

BAEHREL, M.   'La Haine de Classe en temps d'épidémie.' *Annales E.S.C.* t. VII 1952 No 2 p351.

BEAN, J. M. W.   'Plague, Population and Economic Decline in England in the Later Middle Ages.' *Econ. Hist. Rev.* 2nd Series Vol XV 1963, p423.

BERESFORD, M.   *The Lost Villages of England.* London 1954.

*CAMPBELL, A.   *The Black Death and Men of Learning.* New York 1931.

CRAWFURD, R.   *Plague and Pestilence in Literature and Art.* Oxford 1914.

*COHN, N.   *The Pursuit of the Millennium.* London 1957.

D'IRSAY, S.   'The Black Death and the Mediaeval Universities.' *Annals of Medical History.* Vol VII 1925, p220.

D'IRSAY, S.   'Notes to the Origin of the Expression: Atra Mors.' *Isis.* Vol 8 1926, p328.

*D'IRSAY, S.   'Defence Reactions during the Black Death.' *Annals of Medical History.* Vol IX 1927, p169.

FEBVRE, L.   'La Peste Noire de 1348.' *Annales E.S.C.* IV 1949 No 1 p102.

GRAY, H. L.   'The Commutation of Villein Services in England before the Black Death.' *Eng. Hist. Rev.* Vol XXIX 1914, p625.

GUERCHBERG, S.   'La Controverse sur les Prétendus Semeurs de la Peste Noire d'après les traités de peste de l'époque.' *Revue des Études Juives.* Vol 8 1948, p3.

HARE, R.   *Pomp and Pestilence.* London 1954.

HOLMES, G. A.   *The Estates of the Higher Nobility in Fourteenth Century England.* Cambridge 1957.

HOSKINS, W. G.   *The Making of the English Landscape.* London 1955.

JEANSELME, M. E.   'Inondations, Famines et Tremblements de Terre sont les avant-coureurs de la Peste.' *Proc. 3rd Int. Cong. Hist. Med.* 1923.

*KELTER, E.   'Das Deutsche Wirtschaftsleben des 14 and 15 Jahrhunderts in Schatten der Pestepidemien. *Jahrbücher für Nationalökonomie und Statistik.* Vol 165 1953, p161.

KJENNERUD, I. R. 'Black Death.' *Journ. Hist. Med.* Vol 3 1948, p359.

KOSMINSKY, E. A. *Studies in the Agrarian History of England in the 13th Century.* Oxford 1956.

*LANGER, W. L. 'The Next Assignment.' *American Hist. Rev.* Vol LXIII 1958 No 2 p283.

LUTGE, F. 'Das 14/15 Jahrhundert in der Sozial und Wirtschaftsgeschichte.' *Jahrbücher für Nationalökonomie und Statistik.* Vol 162 1950, p161.

*MEISS, M. *Painting in Florence and Siena after the Black Death.* Princeton 1951.

*MICHON, L. A. J. 'Documents inédits sur la Grande Peste de 1348. *Thèses de l'École de Médicine.* Vol VI Paris 1860.

MODE, P. *Influence of the Black Death on English Monasteries.* Chicago 1916.

MOLLARET, H. H. and BROSSOLET, J. *La Peste, Source Méconnue d'Inspiration Artistique.* Paris (Institut Pasteur) 1965.

PAGE, T. W. 'The end of Villainage in England. *Pub. Amer. Econ. Assoc.* Third Series 1900 Vol I.

PARKES, J. *The Jew in the Mediaeval Community.* London 1938.

PERDRIZET, P. *La Peinture Religieuse en Italie jusqu'à la fin du XIV e Siècle.* Nancy 1905.

PERDRIZET, P. *La Vierge de Miséricorde.* Paris 1908.

*POSTAN, M. 'Some Economic Evidence of Declining Population in the Later Middle Ages.' *Econ. Hist. Rev.* 2nd Ser Vol II 1950, p221.

POWER, E. 'The Effects of the Black Death on Rural Organization in England.' *History.* N.S. Vol III 1918, p109.

PUTNAM, B. H. *The Enforcement of the Statute of Labourers.* New York 1908.

*RENOUARD, Y. 'Conséquences et intérêt démographique de la Peste Noire de 1348.' *Population.* t. III 1948, p459.

RÉVILLE, A. *Le Soulèvement des Travailleurs d'Angleterre en 1381.* Introduction by Ch. Petit-Dutaillis. Paris 1898.

ROBBINS, H. 'A Comparison of the Effects of the Black Death on the Economic Organization of France and England.' *Journal of Political Economy.* Vol XXXVI 1928, p447.

ROBERTS, R. S. 'The Place of Plague in English History.' *Proc. Roy. Soc. Med. (Hist. Med.).* Vol 59 1966, p101.

*RUSSELL, J. C. *British Mediaeval Population.* Albuquerque 1948.

RUSSELL, J. S. 'Effects of Pestilence and Plague, 1315-85.' *Comparative Studies in Society and History.* Vol VIII 1966 No 4 p464.

SALTMARSH, J.   'Plague and Economic Decline in England in the Later Middle Ages.' *Camb. Hist. Journ.* Vol VII 1941 No 1 p23.

SIEGRFIED, A.   *Itinéraines des contagions: épidémies et idéologies.* Paris 1960.

THOMPSON, J. W.   'The Aftermath of the Black Death and the Aftermath of the Great War.' *American Journal of Sociology.* 1920/21 Vol XXVI.

THRUPP, S. L.   'The Problem of Replacement Rates in Late Mediaeval English Population.' *Econ. Hist. Rev.* 2nd Series Vol XVIII 1965, p101.

THRUPP, S. L.   'Plague Effects in Mediaeval Europe.' *Comparative Studies in Society and History.* Vol VIII 1966 No 4 p482.

TITOW, J. Z.   'Some Evidence of the 13th Century Population Increase.' *Econ. Hist. Rev.* 2nd Series Vol XIV 1961 No 2 p218.

TONONI, A. G.   'La Peste dell'Anno 1348.' *Giornale Ligustico.* Genoa Vol X 1883, p139.

WILSON, E. M. CARUS   *Mediaeval Merchant Venturers.* London 1954.

The most important contemporary or near-contemporary texts, listed under the country of the chronicler, are as follows:

Italy

Cronica Senese di Agnolo di Tura del Grasso. Muratori. *Rerum Italicarum Scriptores.* 15 VI pp555-57.

Storie Pistoresi. Muratori. 11 V pp235-38.

Gabriel de Mussis. *Historia de Morbo s. Mortalitate quae fuit Anno Dni MCCCXLVII.* ed. Henschel. Haeser's *Archiv. fur die Gesammte Medizin.* II 1842, pp26-59 (cf. V. J. Derbes. 'De Mussis and the Great Plague of 1348.' J.A.M.A. Vol. 196 1966 No 1)

Cronicon Estense. Muratori. 15 III pp159-64.

Michael Platiensis. *Bibliotheca Scriptorum Qui Res In Sicilia Gestas Retulere.* Vol 1 p562.

Lorenzo de Monaci. *Chronicon de Rebus Venetorum.* Venice 1758.

Peter Azarius. Muratori. XVI p16.

Cronica Fiorentina di Marchionne di Coppo Stefani Muratori 30, I.

Cronica de Giovanni Villani. ed. Dragomanni. Florence 1845.

Cronica di Matteo Villani. ed. Dragomanni. Florence 1846.

Monumenta Pisana Muratori. XV (1729 edition) p1021.

Cronicon Gestorum ac Factorum Memorabilium Civitatis Bononie. Muratori. 28 III p43.

Germany

Matthaie Neuwenburgensis. *Cronica*, Fontes Rerum Germanicarum (F.R.G.) ed. Boehmer. Stuttgart 1868, pp261-67. cf. Mon. Germ. Hist. N.S. Vol IV Berlin 1936.

Henricus Dapifer de Diessenhoven, F.R.G. Stuttgart 1868 IV pp68-71.

Hugh von Reutlingen. *Weltchronik.* ed. Gillert. Munich 1881.

Johannis de Winterthur (Vitodurani). Mon. Germ. Hist. N.S. Vol. III.

Annales Engelbergenses. Mon. Germ. Hist. XVII

Kalendarium Zwetlense. Mon. Germ. Hist. IX.

Annales Matseenses. Mon. Germ. Hist. IX.

Henrici de Hervordia. Ed. Potthast. Göttingen 1859.

Continuatio Novimontensis. Mon. Germ. Hist. IX p675.

Heinrici Rebdorfensis Annales Imperatorum. F.R.G. Vol IV pp532-38.

British Isles

Robert of Avesbury. *Continuatio Chronicarum de Gestis Mirabilibus Regis Ed. III.* Ed. E. M. Thompson. R.S. 93.

Thomas Walsingham. *Historia Anglicana.* Ed. H. Riley. R.S. 28.

Polychronicon Ranulphi Higden. Ed. J. R. Lumby. R.S. 41 VIII.

Eulogium (Historiarum sive Temporis). Ed. F. Haydon, R.S. 9 III.

John Capgrave. *The Chronicle of England.* Ed. F. Hingeston, R.S. 1.

Chronicon Henrici Knighton. Ed. J. Lumby, R.S. 92 II.

Gesta Edwardi de Carnavan. Auctori Canonico Bridlingtoniensi cum Continuatione. Ed. W. Stubbs, R.S. 76 II.

Chronicon Galfridi Le Baker de Swynebroke. Ed. E. M. Thompson. Oxford 1889.

Chronicon Johannis de Reading. Ed. J. Tait. Manchester 1914.

Willelmi de Dene. *Historia Rossensis.* Wharton. *Anglia Sacra.* Vol 1 p356. cf. B. Mus. Cotton M.S. Faust B V p96.

Stephen Birchington. *Vitae Archiepiscoporum Cantuariensium.* Wharton, *Anglia Sacra.* Vol 1 p1.

'A Fourteenth Century Chronicle from the Grey Friars at Lynn.' Ed. A. Gransen. *Eng. Hist. Rev.* Vol LXXII 1957, p270.

Chronica Monasterii de Melsa. R.S. 43, III.

Historical Papers from Northern Registers. Ed. J. Raine, R.S. 61.

Eynsham Cartulary. Ed. H. E. Salter. Ox. Hist. Soc. Oxford 1907-8.

Chronicle of Louth Park. Linc. Rec. Soc. 1891.

John Clyn. *Annalium Hiberniae Chronicon.* Ed. R. Butler. Irish Arch. Soc. Dublin 1849.

Chronicle of John of Fordun. Ed. W. F. Skene. Edinburgh 1872.

Cronykil of Andrew of Wyntoun. Ed. D. Laing. Edinburgh 1872.

Book of Pluscarden. Ed. F. J. Skene. Edinburgh 1880.

Other Countries

Continuatio Chronici Guillelmi de Nangiaco. Soc. de L'Histoire de France. t. II 1844, p211.

Chronicon Pragense. Ed. Loserth. Fontes Rerum Austriacarum. Vol I.

Breve Chronicon Clerici Anonymi. De Smet. *Recueil des Chroniques de Flandres*. Vol III p5.

Chronicon Majus Aegidii Li Muisis. De Smet. Ibid. Vol II p110.

C. S. Bartsocas. 'Two 14th Century Greek Descriptions of the Black Death (Nicephoros Gregoras and Emperor John Cantacuzenos).' *Journ. Hist. Med*. Vol XXI 1966 No 4 p394.

Finally there are the plague tractates left by the doctors and savants of the period. Almost all those listed below were analysed by Anna Campbell in her invaluable study *The Black Death and Men of Learning*.

Master Jacme d'Agramont. Sudhoff. *Archiv für Geschichte der Medizin*. XVII (1925) 120-21.

(Klebs, A. C. 'A Catalan Plague Tract of April 24, 1348.' *6ème Congrès International d'Histoire de la Médecine*, Anvers 1929, pp229-232.)

Gentile da Foligno. Sudhoff. *Archiv* V (1913), 83-86, 332-37. A. Philippe. *Histoire de la Peste Noire*. (Paris 1853) contains another text.

John of Penna. Sudhoff. *Archiv* V, (1913), 341-8 and Archiv XVI (1924) pp162-67.

Paris Faculty of the Colleges of Medicine. (*Compendium de Epydimia*). Brit. Mus. Harl. 3,050 (XVII) p66 recto (b) to p68 verso (b). H. E. Rebouis. *Étude historique et critique sur la peste*. Paris 1888. (D. W. Singer. 'Some Plague Tractates.'' *Proc. Roy. Soc. Med*. Vol IX Pt 2 p159.)

Master Albert. Sudhoff. *Archiv* VI, 316-17.

Alfonso de Cordova. Sudhoff. *Archiv* III, 224-26.

Abū Ja'far Ahmad Ibn 'Ali Ibn Mūhammad Ibn 'Ali Ibn Khātimah (referred to generally as Ibn Khātimah). Translated into German in Sudhoff. *Archiv* XIX (1927) by Taha Dinānah. pp27-81.

John Hake of Göttingen. Sudhoff. *Archiv* V, 37-38.

The Five Doctors of Strasbourg (*Treasure of Wisdom and of Art*) Sudhoff, *Archiv* XVI (1924).

(E. Wickersheimer 'La Peste Noire à Strasbourg et le régime des cinq médecins strasbourgeois.' *3ème Congrès International d'Histoire de la Médecine*. Antwerp 1923, pp54-60.)

Author of 'Utrum Mortalitas . . .' ('Is it from Divine Wrath that the Mortality of These Years Proceeds?') Sudhoff. *Archiv* XI, 44-51. Ascribed by S. Guerchberg ('La controverse sur les prétendus semeurs de la Peste Noire' *Revue des Études Juives*. Vol VIII p3 1948) to Konrad de Megenberg.

Abū 'Abdallah Muhammad Ibn 'Abdallāh Ibn Sa'id Ibn Al-Khatīb Lisânal-Din (referred to generally as Ibn al-Khatīb). Translated into German in *Sitzungsberichte der Königl. bayer. Akademie der*

Wissenschaften zu München. II (Munich 1863) 1-28 by M. J. Müller.

Konrad de Megenberg. *Buch der Natur*. Ed. Pfeiffer 1870.

Dionysius Colle (*De pestilentia 1348-1350 et peripneumonia pestilentiali*). D. J. Colle. Benonensi. Pisa 1617, pp570-76.

Author of 'Primo de Epydimia . . .' Sudhoff. *Archiv* V (1913) 41-46.

Simon of Covino. *Bibl. de l'École des Chartes* (1840-41). Ed. E. Littré. Sér 1, Vol 2 pp201-43.

John of Burgundy (John à la Barbe). (Published 1371 but mainly relating to epidemic of 1348-49.) Ed. D. W. Singer. *Proc. Roy. Soc. Med*. Vol 9 Pt 2 p159.

Guy de Chauliac. *La Grande Chirurgie*. Ed. F. Nicaise. Paris 1890.

Index

About the author

About the book

Insights,
Interviews
& More...

Read on

Meet Philip Ziegler

Sophie Ziegler

PHILIP ZIEGLER lives in Kensington, in, or anyway near, the heart of London. He used to have a house in the New Forest, a hundred miles or so to the southwest, but decided reluctantly that, as he grew more decrepit, he must settle for a single home. Children, grandchildren, friends, libraries—all made it inevitable that that home had to be in London. There are plenty of relations and kind friends on whom he and his wife can impose themselves during weekends, and, anyway, when it is possible to walk to the London Library in forty-five minutes, almost entirely through parks, the country seems, if not superfluous, at least dispensable.

He was educated at Eton and New College, Oxford—institutions which seemed the obvious selection at the time but which, in retrospect, suggest an almost embarrassing degree of privilege. From there, after a brief and singularly undistinguished period of national service in the Army, he joined the Foreign Office.

To those who love travel, to be sent abroad at Her Majesty's expense to live in very reasonable comfort and with carte blanche to invade any sector of local society which one finds interesting, is an immensely satisfying way of spending the first fifteen years of a working life. Ziegler had the good fortune to cover his continents—serving in Vientiane, Paris, Pretoria, and Bogotá—but then concluded that the work was unlikely to become noticeably more interesting, that the burdens of diplomatic social life would become more stifling, and that the irritations of traveling around the world with ever-increasing amounts of books, furniture, and children might quickly become intolerable.

By then he had already written two books—biographies of the Duchess of Dino and the early nineteenth-century prime minister Henry Addington—and had also had the good fortune to marry his publisher's daughter. It seemed therefore logical to complete the circle and join his father-in-law's firm, William Collins. He first gave himself six months off in which to start another book.

Mounting the steps which led to the London Library, in search of some ▶

> “ He was educated at Eton and New College, Oxford—institutions which seemed the obvious selection at the time but which, in retrospect, suggest an almost embarrassing degree of privilege. ”

early nineteenth-century figure who had been insufficiently biographized, he met an old friend who asked what he was doing. He explained, and the friend said that he had just been asked to write a book about the Black Death. He had no time to do it himself: Why did not Ziegler take it on? Knowing little about the Middle Ages and still less about epidemiology, Ziegler scoffed at the idea. Out of curiosity, though, he ventured into the part of the library which dealt with plagues and pestilences. Within an hour he had realized that the subject was both enthralling and vastly important, and that the only book which dealt with the subject as a whole was inadequate and out of date.

Four years later, *The Black Death* was published. It had proved possible—just—to combine its writing with a career in publishing, and he had hoped to keep the two in uneasy balance for the rest of his working life. When he was asked to write the official biography of Lord Mountbatten, however, he reluctantly accepted that the demands on his time and energies would be so great that, if only out of fairness towards the authors whose books he was editing, he would have to call it a day. For a middle-aged man with three children, it was a bold decision, but fortunately the publication of *Mountbatten* was followed by an invitation to write the official biography of King Edward VIII, and a measure of security was achieved.

Since then, Ziegler has written

66 Knowing little about the Middle Ages and still less about epidemiology, Ziegler scoffed at the idea. 99

4

biographies of, among others, Osbert Sitwell, Harold Wilson, and the publisher and writer Rupert Hart-Davis; he has also written a description of life in London during the Second World War, and a study of the British soldier based on the mini-biographies of half a dozen Chelsea pensioners. Now age seventy-eight, he is still a full-time writer, at present engaged on the official biography of former prime minister Edward Heath.

Ziegler reads omnivorously, trying usually to follow a novel with a work of nonfiction and a new book with an old one. History, biography, and travel are the nonfiction subjects that he prefers; Stendhal and Jane Austen his favorite novelists from the past; Roth, Updike, and Gabriel García Márquez for the present. No city is richer than London in theatre, opera, and music, all of which he relishes—the problem is to find time to write. ∾

> 66 Ziegler reads omnivorously, trying usually to follow a novel with a work of nonfiction and a new book with an old one. 99

Philip Ziegler on Contemporary Historians and the Black Death

WHEN I FIRST PUBLISHED this book, in 1969, I was apprehensive about the likely reactions of the professional historians. Medievalists are a rugged lot, defending their territory with all the ferocity of a lion who sees its mates threatened by an intruder from another tribe. How would they view this avowed amateur, venturing into areas which have been and still are the scene of so much internecine bickering among the experts? On the whole they treated me with generosity. Their tendency was to treat me not so much as a candidate for demolition as a useful tool with which to assault offending rivals. "Mr. Ziegler describes himself as an amateur," ran one line of argument, "but compared with Professor X of Y University he seems a real professional. . . ." Whether their charity would have endured if I had ventured for a second time into their field, I rather doubt. I did not put their forbearance to the test, but prudently retreated to the nineteenth and twentieth centuries, where I feel rather more at home.

One feature of my book which came in for some criticism was the chapter in which I tried to recapture the impact of the plague in an imaginary medieval village. Such frivolity, it was suggested, was out of place in a book which had any pretensions to scholarship. Historical

> 66 One feature of my book which came in for some criticism was the chapter in which I tried to recapture the impact of the plague in an imaginary medieval village. 99

6

fiction was historical fiction and evermore would be so.

I remain impenitent. Provided an author explains what he is doing and makes no attempt to confuse fiction with fact, surmise with certainty, then I believe he has the right to use any device at his disposal if it will help to make his point more forcibly. In the case of *The Black Death*, I did not feel that it was possible to capture fully the consequences of the plague on a small, credulous, rural society without inventing a village and examining it under the microscope of the imagination. Statistics and facts alone, however striking, could not convey the horror that afflicted Europe in the mid-fourteenth century. My confidence has been reaffirmed by John Hatcher's recent scholarly yet imaginative study of the Black Death in the Sussex village of Walsham le Willows—a book which blends meticulously researched manorial records with colorful detail drawn from sources all over Europe.

In his interesting if somewhat idiosyncratic study of the Black Death and its consequences, Norman Cantor reviewed the existing books on the subject. Mine, he concluded, was "highly readable and out of date." I am delighted to be thought "highly readable" and can return the compliment—Professor Cantor is one of the liveliest and most imaginative of academic historians. "Out of date," in a field in which so much work is going on, is a charge which can hardly be rebutted—even more so now than when Cantor published his book in 2001. Yet when ▶

Philip Ziegler on Contemporary Historians and the Black Death *(continued)*

I review the plethora of books, monographs, and articles in learned journals which have appeared since 1969, I am surprised how few of my conclusions require revision and how many have actually been reinforced.

A great deal of the work has taken place in the medical field, questioning both the nature of the disease and the means by which it traveled from place to place. In my book, I argued that the Black Death in its original form was bubonic plague, which in the course of its travels was made still more lethal and all-pervasive by its modification into pneumonic and septicemic variants of the disease. The bacillus of bubonic plague, *Pasteurella pestis,* found a home in the stomach of a flea—*Xenopsylla cheopis*—and this in turn, when it moved across Europe in the fourteenth century, was carried most often in the hair of the vagabond black rat, *Rattus rattus.*

It is the role of the rat which has been most often questioned. No one, since 1945 at least, has seriously suggested that the rat was the *only* way by which the disease traveled; what is now clear is that in certain cases the rat played no part at all. Gunnar Karlsson, for instance, has pointed out that in Iceland half the population died, yet there were no rats to be found on that island until several centuries later. It is, of course, conceivable that rats ventured ashore from some foreign vessel and managed to discharge their deadly cargo before

> **❝ I am surprised how few of my conclusions require revision and how many have actually been reinforced. ❞**

they perished or returned to their ship, but the speed with which the disease spread suggests some other significant factor must have existed.

Samuel Cohn has similarly shown that the Black Death thrived at times of the year when the fleas from a rat would have had little chance of surviving, while the zoologist, Graham Twigg, has argued that death from the Black Death in England was as likely to occur in thinly as in densely populated areas, and in winter as much as in summer—once again suggesting that *Rattus rattus* was not the exclusive, or even the most significant, villain in the piece.

Such debate assumes that bubonic plague, though made more deadly by its pneumonic and septicemic mutations, was indeed the principal if not the only element in the Black Death. A more radical view is that something quite different was involved: a rare form of the cattle disease anthrax. The symptoms of anthrax, in the initial stages, are consistent with those of bubonic plague, and the speed with which the disease passed from place to place is more characteristic of a cattle-borne disease than one dependent on the slower-moving rodent. Whether anthrax did in fact occur in British cattle during the fourteenth century is difficult to establish; traditionally, it has been believed that it did not, but recent research by Edward Thompson has suggested that there were, indeed, ▶

66 Such debate assumes that bubonic plague . . . was indeed the principal if not the only element in the Black Death. 99

Philip Ziegler on Contemporary
Historians and the Black Death *(continued)*

some outbreaks, and Graham Twigg is convinced that anthrax, or perhaps some other cattle disease, was at least in part responsible for the terrifying mortality. Probably the evidence does not exist on which a final conclusion of these mysteries will ever be achieved; the weight of medical opinion still seems to be that, though other elements were present, both in the disease itself and in the mechanism of its spread, rat-borne bubonic plague was the most significant element in the Black Death.

Debate about the medical background is both interesting and important, but it is the effect of the catastrophe on the people of the time and on succeeding generations which is of more immediate concern to most non-specialist readers. It is still generally accepted that about a third of Europe's population died between 1347 and 1350, but recent research has tended to emphasize the variations in mortality between different parts of the continent. In the Mediterranean area it seems to have been substantially higher than in the British Isles or other northern lands (though Karlsson's figures for Iceland suggest that the reasons for such differences may not be climatic). There are those who see the Black Death as being a Malthusian correction of overpopulation—the continent was outgrowing its resources and nature felt bound to check its overenthusiastic progress. The argument cannot be entirely dismissed, but the fact that mortality was as high or higher in the more thinly populated areas than in the more densely inhabited parts of the continent suggests that nature, if it were indeed seeking to regulate the excesses of humanity, was characteristically casual in the application of its rules.

Much work has been done on the propensity of mankind, when badly frightened by some affliction which it can neither understand nor control, to seek out some group which can plausibly be identified as the cause of the problem and punish it accordingly. The persecution of the Jews during and after the Black Death is the most striking illustration of this and provides the field on which most research has concentrated, but in his interesting study contrasting the effects of the Black Death in Egypt with those in England, Stuart Borsch has pointed out that, in the Middle East, it was women who became the principal scapegoats. The Sultan of Cairo was assured that the Black Death was Allah's punishment for the sin of fornication; and though it can reasonably be contended that it takes two to fornicate, most self-respecting Cairene men of the fourteenth century would have been in no doubt that it was the women who did the tempting and the men who fell.

Much thought too has been given to the long-term social and economic effects of the plague. The propensity of scholars to contradict whatever ▶

> ❝ Stuart Borsch has pointed out that, in the Middle East, it was women who became the principal scapegoats. ❞

theory has been propounded by colleagues from another university is no less marked here than in other fields of academic study, but on the whole the tendency has been to stress, even more strongly than I did, the fundamental changes introduced, or at least accelerated, by the plague. That the Black Death was a contributory cause of the Renaissance and the Reformation has long been taken for granted by most historians; recent studies have accepted this but have emphasized that the pace of change has varied dramatically according to the level of havoc which the plague wreaked. Western Europe, where the death rate was conspicuously higher than further to the east, as a result evolved with far greater speed. Serfdom was virtually a thing of the past in England by the middle of the sixteenth century; in Russia it survived for another three hundred years. As with every other such phenomenon, post hoc is not necessarily propter hoc, but it would seem perverse to deny that the drastic diminution in the size of the labor force was a major factor in ensuring that Western Europe progressed more rapidly than Eastern Europe and that, in due course, the industrial revolution took place in England rather than in Russia.

On the whole I feel that my book on the Black Death has stood up well to the assaults of time and that, for the general reader, it remains the most accessible and comprehensive study of the subject.

It was never intended for medieval scholars, but I hope I am not being overcomplacent when I say that even they would accept that its conclusions are generally sound. At the least, I hope that this book will continue to give pleasure to those who read it and will encourage some to venture further into the history of a cataclysmic event, the effects of which can still be detected in the world we inhabit today. ∽

Author's Picks

The Decameron by Giovanni Boccaccio

As well as being a splendid compendium of often scabrous tales which have echoed down through European literature to the present day, *The Decameron* provides a graphic and moving contemporary account of the impact of the Black Death on Boccaccio's native city, Florence.

A Distant Mirror: The Calamitous Fourteenth Century (1978) by Barbara Tuchman

This book provides a magisterial and brilliantly written background to the Black Death, particularly in France. Tuchman makes the epoch come to life with a vividness and authenticity which should be the envy of any professional medieval historian.

Narcissus and Goldmund (1930) by Hermann Hesse

Narcissus and Goldmund, in which a young monk wanders around a countryside ravaged by the Black Death, captures as forcefully as any twentieth-century novelist could contrive the atmosphere of the century which Barbara Tuchman so rightly called "calamitous."

The Triumph of Death (a fresco) by Andrea Orcagna

One of the few identified major painters who was at work at the time of the Black Death, Orcagna provides in his fresco in the church of Santa Croce in Florence a masterpiece which catches, as powerfully as any other work of art or literature, the stark horror of the impact of the disease at every level of society.

The Triumph of Death, usually featuring a skeletal figure of Death rampant on an equally emaciated horse, is the theme of many other European paintings, for the most part from the early fifteenth century. A particularly fine example, dating from 1445, is to be found in the Galleria Regionale della Sicilia in Palermo. Pieter Bruegel's great *Triumph of Death*, a detail of which appears on the jacket of this book, was painted in 1562 and is in the Prado.

The Seventh Seal (film, 1956) by Ingmar Bergman

Finally, Ingmar Bergman's great film, *The Seventh Seal*, catches the terror of the Black Death as effectively as any book or painting.

The Dance of Death

Hans Holbein

The Young Child

The Dance of Death is also an image much associated with the Black Death, though the earliest known example, in the Cimetière des Innocents in Paris, was not painted until the early fifteenth century. Hans Holbein's marvelously evocative series of woodcuts on the same theme was not issued until 1538.